Stefan Kaiser

Interplay of Charge Order and Superconductivity

Stefan Kaiser

Interplay of Charge Order and Superconductivity

Optical Properties of Quarter-Filled Two-Dimensional Organic Conductors and Superconductors

Südwestdeutscher Verlag für Hochschulschriften

Imprint
Any brand names and product names mentioned in this book are subject to trademark, brand or patent protection and are trademarks or registered trademarks of their respective holders. The use of brand names, product names, common names, trade names, product descriptions etc. even without a particular marking in this work is in no way to be construed to mean that such names may be regarded as unrestricted in respect of trademark and brand protection legislation and could thus be used by anyone.

Publisher:
Südwestdeutscher Verlag für Hochschulschriften
is a trademark of
Dodo Books Indian Ocean Ltd., member of the OmniScriptum S.R.L Publishing group
str. A.Russo 15, of. 61, Chisinau-2068, Republic of Moldova Europe
Printed at: see last page
ISBN: 978-3-8381-2391-2

Zugl. / Approved by: Stuttgart, Uni Stuttgart, Diss. 2010

Copyright © Stefan Kaiser
Copyright © 2011 Dodo Books Indian Ocean Ltd., member of the OmniScriptum S.R.L Publishing group

Abstract

In this thesis the optical properties of quasi two-dimensional organic molecular crystals are investigated. Within this class of materials some systems are metallic, some are insulating due to charge-order, and others even become superconducting. The driving force behind the ground states is the different degree of effective electronic correlations. Recent experimental and theoretical studies of strongly correlated materials suggest that fluctuations of the ordered state may mediate superconductivity. The intention of this work is to prove the presence of such charge fluctuations and reveal their relation to superconductivity. To do this, we investigated BEDT-TTF based organic conductors by optical spectroscopy covering a broad frequency (8-20000 cm^{-1}) and temperature (1.8-300 K) range using FTIR- and THz-spectrometers. In addition, the transport properties are investigated by dc four-point contact measurements and microwave cavity perturbation technique.

We found that in-plane optical conductivity of the superconductor β''-(BEDT-TTF)$_2$-SF$_5$CH$_2$CF$_2$SO$_3$ (T$_c$=5.4 K) shows a small Drude-response, a charge-transfer band, and a charge fluctuation band that represent the interplay of the coherent and localized system. A collective charge-order excitation is present as low-lying lattice phonon. Microwave transport properties also show first traces of charge fluctuations in the superconductor: Resistivity increases right above the superconducting transition temperature while dc properties stay metallic. In the optical spectra below the superconducting transition temperature the superconducting gap is observed. The size of the gap $2\Delta(0) \approx 12$ cm^{-1} suggests a weak coupling with $2\Delta/k_BT_c \approx 3.3$. The gap can be fitted by BCS theory. It can be closed by applying a critical magnetic field of 3.4 T. In contrast, the optical response of the isostructural metal β''-(BEDT-TTF)$_2$SF$_5$CHFSO$_3$ shows a strong Drude response and only a small charge-transfer band.
Out-of-plane vibrational studies allow a characterization of the charge-order in the systems by measuring frequencies of charge sensitive vibrations of the BEDT-TTF molecule. They split on charge-order and the frequency splitting reveals the charge redistribution between the molecular sites. For insulators charge difference between the sites is above $0.6e$. Interestingly, also the superconductors show a splitting which corresponds to a significant charge disproportionantion between the sites of about $0.2e$.

Summarizing, the systems show superconductivity in proximity to a charge-ordered state. The superconductors show strong signatures of charge fluctuations due to high effective electronic correlations. That reveals the strong relation between the charge-order and superconductivity. The metals are less correlated. Fluctuations of charge order are absent or weak.

The experimental results are compared to a theoretical model of a so-called charge-ordered metallic state. Within this model charge-fluctuations close to the charge-order transition mediate the attractive interaction between the charge carriers that leads to the formation of Cooper-pairs at low temperature. A comparison with other correlated systems shows that the low frequency properties are similar to the bad-metal regime of high-temperature superconductors. Further, the excitations of the charge-order electronically or via a lattice phonon can be described as collective mode of the ordered state. Besides the electronic interactions, a strong coupling to lattice dynamics is found. First attempts are made to include this into the picture of electronic correlations.

Contents

1. **Introduction**   7

I. Correlated electron systems   11

2. **Organic conductors as correlated electron systems**   13
   2.1. Physics of correlated electron systems
   and competing ground states 13
        2.1.1. The Drude Lorentz model 13
        2.1.2. The Fermi liquid and highly correlated electron systems 16
        2.1.3. The Hubbard model and the extended Hubbard model 19
        2.1.4. Optical properties of organic conductors 19
   2.2. Low-dim. organic conductors as model system for correlated electron systems 21
        2.2.1. Structure and electronic system 21
        2.2.2. Influence of electronic correlations 26
        2.2.3. Optical properties of half-filled systems 27
        2.2.4. Optical properties of quarter-filled systems 30
        2.2.5. One fifth-filled systems 39

3. **Interplay of charge-order and superconductivity**   43
   3.1. Description of 2D quarter-filled systems using the extended Hubbard model   43
        3.1.1. Nearest neighbor interaction driven charge order 45
        3.1.2. Superconductivity mediated by charge fluctuation 46
   3.2. Dynamics and spectral traces of metal, charge order, and superconductivity   47
        3.2.1. Metallic system 48
        3.2.2. Charge order 49
        3.2.3. Dynamical properties of the charge ordered metal 49
        3.2.4. Superconductivity 51
   3.3. Aspects of vibrational spectroscopy 54
        3.3.1. Charge dependent molecular vibrations 55
        3.3.2. Electron molecular vibration coupling 56

4. **Investigated organic systems**   59
   4.1. The β''-(BEDT-TTF)$_2$SF$_5R$SO$_3$ family 59
        4.1.1. β''-(BEDT-TTF)$_2$SF$_5$CH$_2$CF$_2$SO$_3$ 60

Contents

	4.1.2. β''-(BEDT-TTF)$_2$SF$_5$CHFSO$_3$	71
4.2.	The β-(EDT-TTF)$_4$[Hg$_3$I$_8$]$_{(1-x)}$ organic superconductor	76
4.3.	The α-(BEDT-TTF)$_2$$M$Hg(SCN)$_4$ family family	79
4.4.	The θ-(BEDT-TTF)$_2$RbZn(SCN)$_4$ and α-(BEDT-TTF)$_2$I$_3$ charge order compounds	80
	4.4.1. α-(BEDT-TTF)$_2$I$_3$	81
	4.4.2. θ-(BEDT-TTF)$_2$RbZn(SCN)$_4$	81

5. Aims of the thesis: Interplay of CO and superconductivity — **85**
  5.1. Characterizing the degree of charge order in a metallic, charge ordered and superconducting phase 85
  5.2. Influence of the charge order to the optical properties 86
  5.3. Overview picture about the interplay of charge fluctuations and superconductivity 86

II. Experiments 87

6. Experimental techniques **89**
  6.1. Optical measurements 89
    6.1.1. Broadband FTIR spectroscopy in the VIS/NIR-FIR range 89
    6.1.2. THz spectroscopy 92
  6.2. Vibrational spectroscopy 94
    6.2.1. Raman microscopy 96
    6.2.2. IR microscopy 97
  6.3. Optical low temperature systems and sample mounting 98
    6.3.1. Gas exchange cryostat with sliding aperture technique 98
    6.3.2. Cold finger cryostat for gold evaporation 98
    6.3.3. Micro cryostat with micro positioning unit for IR microscopy . . 99
  6.4. Transport measurements 100
    6.4.1. Four and two probe contact method for DC conductivity 100
    6.4.2. Microwave cavity perturbation 100

7. Data analysis **101**
  7.1. Cleaning and merging data 101
    7.1.1. Atmospheric and beam-splitter artifacts 101
    7.1.2. Merging the data 102
  7.2. Calculating the optical response and data fit 103
    7.2.1. Drude Lorentz fit 104
    7.2.2. Extended Drude formalism 104

III. Results 105

8. Transport measurements — 107
8.1. The organic superconductor $\beta''\text{-}(BEDT\text{-}TTF)_2SF_5CH_2CF_2SO_3$ 107
8.2. The deuterated superconductor $\beta''\text{-}(d_8\text{-}BEDT\text{-}TTF)_2SF_5CH_2CF_2SO_3$ . . 110
8.3. The organic metal $\beta''\text{-}(BEDT\text{-}TTF)_2SF_5CHFSO_3$ 110
8.4. The EDT based superconductor $\beta\text{-}(EDT\text{-}TTF)_4[Hg_3I_8]_{(1-x)}$ 114
8.5. The metal insulator transition in $\alpha\text{-}(BEDT\text{-}TTF)_2I_3$ 115
8.6. Summary of transport properties 115

9. Vibrational spectroscopy — 117
9.1. On-site charge distribution in the organic superconductors and metals . . 117
    9.1.1. The organic superconductor $\beta''\text{-}(BEDT\text{-}TTF)_2SF_5CH_2CF_2SO_3$. 117
    9.1.2. The $\beta''\text{-}(BEDT\text{-}TTF)_2SF_5CHFSO_3$ organic metal 125
    9.1.3. The $\beta\text{-}(EDT\text{-}TTF)_4[Hg_3I_8]_{(1-x)}$ organic superconductor 131
    9.1.4. The $\alpha\text{-}(BEDT\text{-}TTF)_2NH_4Hg(SCN)_4$ and $\alpha\text{-}(BEDT\text{-}TTF)_2TlHg(SCN)_4$ organic metals 136
9.2. The metal insulator transition in $\alpha\text{-}(BEDT\text{-}TTF)_2I_3$ and $\theta\text{-}(BEDT\text{-}TTF)_2$-$RbZn(SCN)_4$ 140
    9.2.1. The charge order transition in $\theta\text{-}(BEDT\text{-}TTF)_2RbZn(SCN)_4$ . . . 141
    9.2.2. The metal insulator transition in the $h8\text{-}\alpha\text{-}(BEDT\text{-}TTF)_2I_3$ and $d8\text{-}\alpha\text{-}(BEDT\text{-}TTF)_2I_3$ system 145
9.3. Charge Order and charge redistribution for metals, superconductors, and insulators 153
9.4. Properties of the in-plane emv-coupled modes 156

10. Broad band optical measurements — 159
10.1. The in plane response of the $\beta''\text{-}(BEDT\text{-}TTF)_2SF_5CH_2CF_2SO_3$ superconductor and the isostructural $\beta''\text{-}(BEDT\text{-}TTF)_2SF_5CHFSO_3$ metal . . 159
10.2. Temperature dependence of the in-plane optical response 162
    10.2.1. $\beta''\text{-}(BEDT\text{-}TTF)_2SF_5CH_2CF_2SO_3$ 162
    10.2.2. $\beta''\text{-}(BEDT\text{-}TTF)_2SF_5CHFSO_3$ 179
    10.2.3. Comparison 183
10.3. The superconductor $\beta\text{-}(EDT\text{-}TTF)_4[Hg_3I_8]_{(1-x)}$ 185
10.4. THz spectroscopy of the superconducting gap 190
    10.4.1. Absolute reflectivities in the superconducting gap regime 191
    10.4.2. Opening of the superconducting gap 193
    10.4.3. Comparison to the superconducting gap in $\alpha_t\text{-}(BEDT\text{-}TTF)_2I_3$ . . 195

IV. Discussion — 199

11. The Interplay of Charge Order and Superconductivity — 201
11.1. The charge ordered metallic state and superconductivity 201
    11.1.1. Transport properties and the low-frequency response of bad metals 201
    11.1.2. The in-plane optical response of the bad metal state 203

11.2. The charge fluctuation interaction for different compounds 207
11.3. Lattice phonon and collective charge order excitation 209
11.4. Vibrational influence to superconductivity 214

V. Summary and Outlook 217

12. Summary 219

13. Outlook 225

VI. Appendix and Bibliography 227

14. Appendix 229
14.1. Slave boson treatment on the extended Hubbard model 229
    14.1.1. Nearest neighbor interaction driven charge order 229
14.2. Construction sketches of the sample holder 232

Bibliography 257

1. Introduction

Organic conductors are a novel class of materials showing intriguing new physics and serving as ideal model compounds for the investigation of low dimensional physics and correlated electron systems [1–3]. Starting from the building blocks of single molecules a large variety of molecular crystals can be synthesized making this class of systems very versatile with a lot of possibilities for modifications. Besides using different sets of donor molecules in combination with various atomic or molecular anions also the change of the packing pattern offers the opportunity for tailored matter by directly engineering electronic properties like band structure, band filling, and the influence of electronic correlations.

Electronic correlations are shown to be the driving force behind a lot of the fascinating and unconventional properties within a large family of correlated systems. The origin of correlated electrons can be very different, e.g. in heavy fermions, metallic compounds where correlations are due to a hybridization of s- and f-electrons [4] while in transition metal oxides, where the cuprate high temperature superconductors are the most prominent example [5, 6], the d-electrons in the covalent bonds are correlated [7]. The correlation in the organics, which are in the focus of the thesis, occurs between the delocalized electrons of the molecular π-orbitals forming the conduction band. However, all these different families of the correlated electron systems show some common behavior and properties [8]. A general feature is the presence of different electronic phases and ground states. These are tuned by the degree of effective electronic correlations in the systems. The class of organic organic conductors for instance shows renormalized metallic behavior for low, and insulating ordered phases in case of high correlations. At intermediate correlations even superconductivity can be found. Among different correlated systems, the superconducting state is often found close to an ordered phase. Then the origin of superconductivity can be discussed in the framework of the interplay and competition of the different ground states. Order fluctuations are suggested to destabilizes the metallic phase towards the ordered phase at a quantum phase transition. These fluctuations could even mediate superconductivity. They interact with the metallic system resulting in an attractive potential for the interacting charge carriers [9, 10]. The nature of the ordered state can be of various manner: ferromagnetic-, or antiferromagnetic order as e.g. in heavy fermions; a density wave state (charge or spin) in transition metal oxides, or a charge ordered state like in the present case of the organic systems. Understanding the mechanism in one system in detail also helps to draw analogies to the behavior and interactions in the other systems. Further, understanding deviations from the proposed general behavior helps to open a way to establish an unified view among the different systems.

1. Introduction

In the case of organic systems the research aims in several directions because of their versatility and easy tuning possibilities of different intrinsic or external parameters (dimensionality, filling, doping, pressure, chemical pressure, external fields, etc.). Main branches are one- or two-dimensional systems at different filling: half-, or quarter-filled but also incommensurate filling like fifth-filled systems.
Considerable knowledge about organics is based on the one-dimensional Bechgaard-Fabre salts [11–13]. They show a large variety of structural and correlation driven phase transitions. Also dimensionality tuning from a quasi one-dimensional systems to two dimensions or even approaching three-dimensional behavior was shown. Basic questions on Mott insulators, Luttinger liquids, and transitions to Fermi liquids are investigated [14]. In contrast to the one-dimensional systems for two-dimensional materials no generic phase diagram exists. This is because different crystal structures and filling of the conduction band leads to different ground states [15, 16]. In the two dimensional case a strong focus is on half-filled compounds because of their close relationship to the cuprate high-temperature superconductors [8]. As the cuprates, κ-BEDT-TTF salts show an antiferromagnetic phase close to the superconducting state. The pressure dependent phase diagram of the organics (where hydrostatic pressure or chemical pressure tunes the bandwidth and thus the degree of effective correlation) looks similar to the cuprates phase diagram if the pressure dependence is replaced by doping. Further, the organics show similar unconventional behavior in the metallic phase. Topics of interest in these and related compounds are Mott-Hubbard transitions, or superconductivity mediated by spin fluctuations. For the organics the physics is driven by the ratio of the electron-electron interaction to the bandwidth of the conducting charge carriers U/t. These systems can be described with the Hubbard model $H = T + U$. The kinetic term T accounts for the hopping probability of charge carriers from (molecular) lattice site to lattice site. It is given by the overlap of the orbitals (represented by the overlap- or hopping integral t) which determines the bandwidth of the system. U describes the correlation between the charge carriers on the same site due to the Coulomb repulsion.

The two-dimensional quarter-filled organic conductors discussed in this thesis also show a peculiar phase diagram. Their ground state depends on the effective inter-site correlations V. Experiments evidence a superconducting state in proximity to an insulating charge-ordered phase. The charges in the system form order patterns of high and low occupied sites due to the electron-electron repulsion and no free transport is possible. This state cannot be understood within the simple Hubbard model taking into account only the on-site repulsion U. Based on experimental findings, calculations on the extended Hubbard model $H = T + U + V$ are performed to describe the dynamical properties of the system. They propose that the insulating state is formed due to large values of the effective on-site and inter-site Coulomb repulsion U/t and V/t, respectively. For weak correlations the compounds are metallic. But close to the metal-insulator phase boundary, charge fluctuations are observed, while the response of coherent carriers is still present. Superconductivity is predicted on the border between these metallic and insulating states. Charge-order fluctuations were proposed to act as an attractive interaction of quasi-particles to form Cooper pairs in the superconducting state [17–19]. While the charge-ordered insulating state is quite well investigated, the region close to the charge-order transition, where the influence of the charge fluctuations is strong, is not fully understood. Also one question is not answered, yet: Whether there is an interplay of the different ground states, as proposed by theory, or more a competition as known from charge-density wave and superconductivity, or static stripes in cuprates, for example. Therefore metallic, insulating, and superconducting organic systems are investigated in this thesis. The superconductors are proposed to be in the vicinity to the charge-order transition. Optical spectroscopy and transport measurements are performed to gain insight into the interplay of the different states that lead to or compete with superconductivity.

Therefore infrared (IR) reflectance is measured in a broad frequency range (5-10000 cm^{-1}) using coherent wave THz- and Fourier Transform Infrared (FTIR) spectrometers in a temperature range from 1.8-300 K. That probes the properties of the metallic and localized charge carriers. In addition it traces the interaction between localized and itinerant carriers. It appears as an optical feature at low but finite frequencies. At lowest temperatures the opening of the superconducting gap is investigated.
Vibrational spectroscopy (Raman and IR) allows one to trace the coupling of molecular and lattice vibrations to the electronic background. But even more important it gives a tool to directly measure the charge located on the molecular sites. Applying these methods to the organic systems gives a tool to investigate the charge redistribution within the system. That proves a charge-order in insulating systems and the existence of charge-order fluctuations in superconductors.

The thesis is organized as follows: Part I sets the basics of physics and materials. First an introduction is given to two-dimensional organic conductors as correlated electron systems in Sec. 2. It is presented how the structure influences the electronic band and what is the influence of the electronic correlations. The compound in the focus of this study, the organic superconductor β''-(BEDT-TTF)$_2$SF$_5$CH$_2$CF$_2$SO$_3$, is introduced in Sec. 4 as an ideal model system to investigate the interplay of charge order and superconductivity. Then related materials are introduced, a study of which helps to clarify dependences on different correlation, influence of the lattice, etc. In Sec. 3 a theoretical model based on the extended Hubbard model is presented that tries to cover the essential physics and helps to explain the observed features. Further it proposes a phase diagram for interacting electrons in a quarter-filled two-dimensional system. Part II gives an overview on the experimental techniques and the data treatment. The results on the investigated compounds are described and discussed in Part III. Part IV draws the conclusions for the class of quarter-filled organic conductors but also for the mechanism of superconductivity mediated by charge fluctuations in general. It compares the features also to other correlated electron systems to show the general nature of this model.

Part I.

Correlated electron systems

Part 1

2. Organic conductors as correlated electron systems

The physics of correlated electron systems is rich of fascinating and intriguing new phenomena in different types of materials. All these various systems have in common that the driving force behind their physical properties is their strong electron-electron interaction. This gives rise to a variety of interesting ground states ranging from insulating charged ordered states, different magnetic exchange phases, metallic or bad metal phases to superconductivity. Special interest in this thesis is given to the superconductivity in organic molecular crystals and in particular to the interplay of the metallic state and charge order phenomena in quarter-filled quasi-two-dimensional systems. In these compounds the superconductivity maybe mediated by charge fluctuations. Therefore, first an overview is given to the physics in highly correlated electron systems in general. Then the subclass of low dimensional organic conductors is presented and a motivation is given why these are the ideal model system to investigate the aforementioned physics.

2.1. Physics of correlated electron systems and competing ground states

2.1.1. The Drude Lorentz model

The simplest model to describe electronic properties in solids is the Drude model [20–25]. It explains the electron dynamics of most metals in a reasonable way. In that case the electrons are described as a non-interacting free electron gas. Their average velocity v is given by their temperature T, and the velocity vector is arbitrary in direction and just changed by elastic scattering. The scattering takes place at the defects of the crystal lattice or phonons only and leads to thermal equilibrium with the surrounding within the average relaxation time τ. Two parameters describe the physics in these systems. One is the plasma frequency $\omega_p = \left(\frac{4\pi N e^2}{m}\right)^{1/2}$, the eigen- or resonance frequency of the quasi-free electron gas, where N is the density of electrons and m the free electron mass. The Drude model also is used to describe simple metals. In that case the material parameters are read as the density of free electrons N and the electron band mass m, which represents all lattice interactions of the electrons and the solids band structure. Even electronic interactions can be taken into account. In that case the electron-electron interactions are put into an effective electron mass. The second parameter is the scattering rate

2. Organic conductors as correlated electron systems

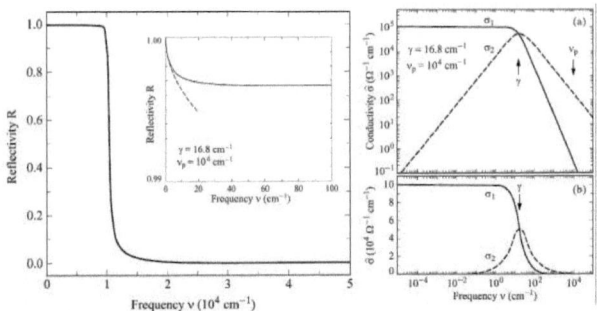

Figure 2.1.: Optical response of a Drude peak with a plasma frequency of $\nu_p = \omega_p/2\pi c = 10000$ cm^{-1} and scattering rate of $\gamma = 16.8$ cm^{-1}. In reflectivity (left) the drop at the plasma frequency is seen where the system becomes transparent. At low frequencies the square root dependence in the Hagen Rubens regime is indicated as dashed line in the inset. The optical conductivity (right) is given on a logarithmic (top) and linear (bottom) scale vs. a logarithmic frequency scale. It shows the constant conductivity in the low frequency range with the roll off starting at the scattering rate. From [22].

$\gamma = \frac{1}{\tau}$, which describes how often the electrons scatter per second. Drude derived, with $l = v\tau$ the mean free path between two scattering events, the conductivity to $\sigma_{dc} = \frac{e^2}{Nm} \frac{l}{v} = \frac{e^2}{Nm}\tau$. In quantum mechanical treatment the electrons have to obey the Fermi distribution. Instead of the thermal velocity, the Fermi velocity is important and only electrons in the $k_B T$ shell around the Fermi energy can be excited. The frequency dependent conductivity in the Drude model [22] results in

$$\sigma(\omega) = \frac{ne^2}{m}\frac{1}{\gamma - i\omega} = \frac{\omega_p^2}{4\pi}\frac{1 - i\omega\tau}{1 + \omega^2\tau^2} = \frac{\omega_p^2}{4\pi}\frac{1}{\frac{1}{\tau} - i\omega} = \frac{\sigma_{dc}}{1 - i\omega\tau}. \tag{2.1}$$

The sum rule integrating the area under this Drude peak is connected to the plasma frequency:

$$\int_0^\infty \sigma_1(\omega)d\omega = \frac{\pi Ne^2}{2m} = \frac{\omega_p^2}{8} \tag{2.2}$$

which measures the number of electrons assuming a fixed electron mass. Thus it is also a measure for the influence of correlation in the system since for a fixed or known number of electrons it measures the effective mass.

2.1. Physics of correlated electron systems and competing ground states

Optical properties of the Drude Lorentz model

The optical response of a metallic system can be described based on the Drude model (Eqn. 2.1) [22]. The zero-frequency Drude peak (Fig. 2.1) of the model can be derived in three regimes. One is the Hagen-Rubens regime describing the frequency range where the transport properties are dominated by the dc conductivity of the system. The $(\omega\tau)$ term can be neglected. The conductivity is basically frequency independent $\sigma(\omega) = \sigma_{dc}$ and the reflectivity calculated from that can be expressed by the Hagen-Rubens relation

$$R(\omega) \approx 1 - \left(\frac{2\omega}{\pi\sigma_{dc}}\right)^{\frac{1}{2}}. \tag{2.3}$$

The frequency range between scattering rate $\gamma = \frac{1}{\tau}$ and plasma frequency ω_p describes the relaxation regime. The $(\omega\tau)$ term cannot be neglected anymore. The conductivity starts to become frequency dependent and the roll off of the Drude peak takes place with approximately a $\frac{1}{(\omega\tau)^2}$ behavior. The high-frequency region is characterized by a sharp drop in reflectivity at the plasma edge and the material becomes transparent well above the plasma frequency. The conductivity in that regime shows no indications of the plasma frequency and continues to drop with a ω^{-2} behavior. In case of a poor conductor the Drude peak is over-damped, the plasma frequency is smaller than the scattering rate $\omega_p < 1/\tau$.

For semiconductors the Lorentz model is a proper phenomenological description for electronic properties due to the semiconducting gap [22]. It is based on a harmonic oscillator $\ddot{\mathbf{r}} + \frac{1}{\tau}\dot{\mathbf{r}} + \omega_0^2 \mathbf{r} = -\frac{e}{m}\mathbf{E}(t)$ at a center frequency ω_0. It describes the development of an energy gap at low frequencies and is widely used to explain non-conducting materials. The Drude model can be considered as zero frequency oscillator since there is no energy gap to excite above. In fact a Lorentz oscillator at $\omega_0 = 0$ reduces to the Drude response. The Lorentz oscillators

$$\sigma(\omega) = \sigma_1(\omega) + i\sigma_2(\omega) = \frac{\omega_p^2}{4\pi}\frac{\omega^2\tau}{(\omega_0^2 - \omega^2)^2 + \omega^2\tau^2} + i\frac{\omega_p^2}{4\pi}\frac{\omega(\omega_0^2 - \omega^2)}{(\omega_0^2 - \omega^2)^2 + \omega^2\tau^2} \tag{2.4}$$

are used to describe inter-band transitions or correlation- and dimerization-gaps but also for contributions of phonons or molecular vibrations. In the case of phonons the damping $1/\tau$ represents the broadening of the oscillator while the oscillator strength is given by the 'plasma frequency' $\omega_p = (4\pi Ne^2/m)^{1/2}$. In the case of electronic bands the model is also used to describe the broad and often over-damped bands showing up in the mid- (MIR) and near-infrared (NIR) region due to electronic correlations in the systems. It is used as a phenomenological model to describe an excitations above a gap not taking into account the specific band structure and electronic density of states or transition matrix elements. But it is sufficient to trace changes and transfer of spectral weight, frequency shifts, etc. of the mimicked bands with temperature or at a phase transition. In the optical properties the Lorentz oscillator (Fig. 2.2) shows a small conductivity and only little absorption in its low frequency range. The dielectric constant saturates

2. Organic conductors as correlated electron systems

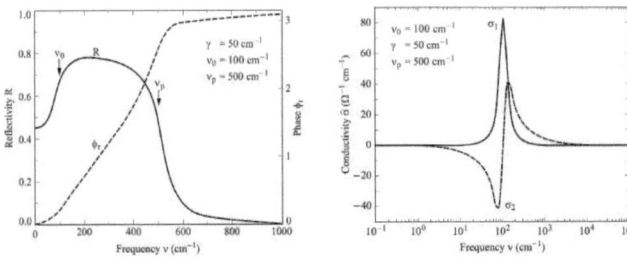

Figure 2.2.: Optical response of the Lorentz oscillator placed at $\nu_0 = \omega_0/2\pi c = 100$ cm^{-1} with an oscillator strength of $\nu_p = \omega_p/2\pi c = 500$ cm^{-1} and width of $\gamma = 50$ cm^{-1}. The reflectivity (left) increases at the center frequency and drops at the 'plasma frequency'. The optical conductivity (right) shows the Lorentz peak located at the center frequency. From [22].

at low frequencies and determines the level of reflectivity. The excitation of electrons above the gap sets in at the center frequency: The conductivity is high and reflectivity increases. The area of this regime is spread by the damping rate around the center frequency $\omega_0 \pm 1/2\tau$. Above the reflection regime there is the area of basically frequency independent reflectivity. The drop of reflectivity takes place around the plasma frequency entering the transparent regime at high frequencies.

2.1.2. The Fermi liquid and highly correlated electron systems

Taking into account electron-electron interactions leads to some completely different effects. The simple ones are screening effects, meaning the real properties of a single electron are hidden by other electrons or charges. In this case one deals with the screened properties of electrons. But as soon as the strength of the electronic correlations is larger than the screening energy, or the electrons are closer together than the screening length, the single electrons really 'feel' their nearest neighbors or even next nearest neighbors. Many-body effects become important especially in cases of a narrow bandwidth in the energy picture or a reduced dimensionality in the picture of screening-length-scale. Then the electron density is enhanced. That is why electron-electron correlations often become important in low-dimensional conductors, etc. Landau introduced the concept of renormalization for interacting fermions [26] as a general model. The interacting electron system can be described as a system of interaction-free quasi-particles, the Fermi liquid. Starting from a free electron gas, the adiabatic on-turn of interactions between the electrons leads to a distortion of the free state with a cloud of interacting electrons. They form a dressed state. In that case, the interactions lead to the renormalization of the dynamical properties (velocity, effective mass, scattering rate, and lifetime due to single particle excitations) which are describing the states. But their basic description is

2.1. Physics of correlated electron systems and competing ground states

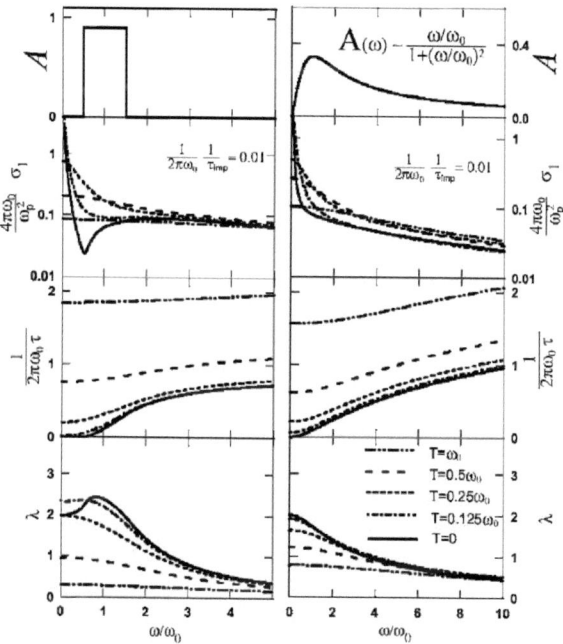

Figure 2.3.: Model calculations of the optical conductivity σ_1, scattering rate $1/\tau$, and mass enhancement λ for interaction with a bosonic spectral function A. A square spectrum in the left panels and a typical spin fluctuations spectrum in the right panels. From [23].

still a Drude response just with reduced spectral weight caused by the enhanced effective mass.

Extended Drude model

To treat the influence of electronic correlation on the optical response, the extended Drude description can be used. It generalizes the standard Drude model by taking a frequency dependent scattering into account. That is done by writing the damping term in the Drude formula as a complex memory function $1/\tau = M(\omega) = M'(\omega) + iM''(\omega)$ [21, 23, 24]. Therefore the complex conductivity can be expressed using the memory

2. Organic conductors as correlated electron systems

function $M(\omega) = 1/\tau(\omega) - i\omega\lambda(\omega)$:

$$\sigma(\omega) = \frac{1}{4\pi} \frac{\omega_p^2}{M(\omega) - i\omega} = \frac{1}{4\pi} \frac{\omega_p^2}{1/\tau(\omega) - i\omega[1 + \lambda(\omega)]}, \qquad (2.5)$$

where $1/\tau(\omega)$ describes the frequency-dependent scattering rate and $\frac{m^*(\omega)}{m} = 1 + \lambda(\omega)$ the mass enhancement of the electronic system. That shows the influence of the many-body interactions that couple the quasi-particle excitations of the Fermi liquid to an (in the present case bosonic) excitation spectrum. Introducing a renormalized scattering rate $1/\tau^*(\omega) = 1/\tau(\omega)[1 + \lambda(\omega)]$ and an effective plasma frequency $\omega_p^{*2}(\omega) = \omega_p^2(\omega)[1 + \lambda(\omega)]$ reduces the conductivity

$$\sigma(\omega) = \frac{1}{4\pi} \frac{\omega_p^{*2}}{1/\tau^*(\omega) - i\omega} \qquad (2.6)$$

formally to the simple Drude equation (Eqn. 2.1). In that case $1/\tau^*(\omega)$ can be seen as a width of a Drude peak local to the finite frequency ω, and $\lambda(\omega)$ as an interaction induced velocity renormalization of the electrons on the Fermi surface [24]. The renormalized scattering rate $1/\tau^*$ is not causal and does not have a physical meaning. It is affected by lifetime effects and the velocity renormalization λ: $1/\tau^*$ and λ are not Kramers Kronig consistent. However the non-renormalized scattering rate $1/\tau$ and the mass renormalization $\frac{m^*(\omega)}{m} = 1 + \lambda$ are causal and can be obtained from the complex conductivity as

$$\frac{1}{\tau(\omega)} = \frac{\omega_p^2}{4\pi} \frac{\sigma_1(\omega)}{|\sigma(\omega)|^2} \qquad (2.7)$$

and

$$\frac{m^*(\omega)}{m} = \frac{\omega_p^2}{4\pi} \frac{1}{\omega} \frac{\sigma_2(\omega)}{|\sigma(\omega)|^2}. \qquad (2.8)$$

For a Fermi liquid the scattering rate is expected to behave quadratic as a function of frequency as well as a function of temperature:

$$\frac{1}{\tau(\omega, T)} = \frac{1}{\tau_0} + a(\hbar\omega)^2 + b(k_b T)^2, \qquad (2.9)$$

where $1/\tau_0$ is impurity scattering resulting in a frequency independent background [21–24]. As described in Sec. 2.1.1, for a simple Drude metal the presence of these impurity scattering is expected, only. The electron-electron interactions in the Fermi liquid result in a square dependence. In case of strong coupling to bosonic features, a linear frequency dependence of the scattering rate is expected [23, 24, 27] as sketched in Fig. 2.3 for two model cases. The respective mass enhancement describes the boson cloud that dresses and is dragged by the charge carriers.

2.1. Physics of correlated electron systems and competing ground states

2.1.3. The Hubbard model and the extended Hubbard model

In some cases the important facts of real systems can be modeled by the Hubbard model. It works well in 1D and in 2D systems. It is a discrete model good for nearly localized systems. Sometimes on can reduce it to on-site and nearest-neighbor interaction and still catch the important physics well. In rare cases of $\frac{1}{2}$-filled systems the main physics is mimicked by only taking into account the on-site interaction U [8]. Then the Pauli principle regulates that only two electrons of opposite spin are allowed to occupy the same site. But even in this case one has to overcome the Coulomb repulsion between these electrons due to their electronic charge. The Hubbard model takes into account the hopping properties of the electrons. They are given by the transfer or overlap integral t which represents their kinetic energy and describes the bandwidth and delocalization of the non-interacting electrons. Electronic correlations are introduced by the Coulomb repulsion between two electrons on the same site U:

$$H = t \sum_{\langle ij \rangle, \sigma} \left(c_{i\sigma}^{\dagger} c_{j\sigma} + c_{j\sigma}^{\dagger} c_{i\sigma} \right) + U \sum_{i} n_{i\uparrow} n_{i\downarrow}, \qquad (2.10)$$

where $c_{i\sigma}^{\dagger}$ is the creation and $c_{i\sigma}$ the corresponding annihilation operator for an electron at site i, and spin σ. n is the electron number operator. Within the model it is possible to investigate metallic or insulating behavior, metal-insulator transition, and superconductivity depending on t, U, band filling, and dimension [28, 29].

In the two-dimensional quarter-filled case the sites are occupied just by half an electron in average. There are always empty neighboring sites for the electron to hop to. Therefore it is important to take into account the intersite Coulomb repulsion V which could be neglected in the half-filled case. This is done in the extended Hubbard model [15]:

$$H = t \sum_{\langle ij \rangle, \sigma} \left(c_{i\sigma}^{\dagger} c_{j\sigma} + c_{j\sigma}^{\dagger} c_{i\sigma} \right) + U \sum_{i} n_{i\uparrow} n_{i\downarrow} + V \sum_{\langle ij \rangle} n_i n_j. \qquad (2.11)$$

The Hubbard model and the extended Hubbard model are widely used to describe all kind of correlated electron systems. It can be also extended to a multi-band Hubbard model. In the case of molecular solids, usually the HOMO-LUMO approach for the bands is sufficient as explained in the following Sec. 2.2. Often an anisotropy of the lattice is present. It can be fully taken into account by the transfer integrals t_{ij} and Coulomb repulsion between the sites V_{ij} if these parameters are taken into the sums.

2.1.4. Optical properties of organic conductors

The influence of the electronic correlations is seen in the optical properties of the organic conductors. They show contributions due to conducting electrons, phonons and interband transitions. A spectrum, as typically found in organic conductors, is shown in Fig. 2.4 sketching the different spectral contributions.

As introduced in Sec. 2.1.1 the free carrier response of simple metals is described within

2. Organic conductors as correlated electron systems

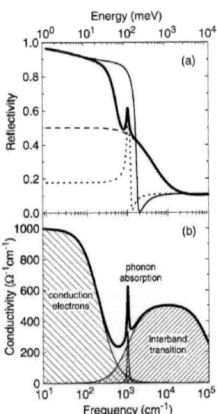

Figure 2.4.: Model of (a) reflectivity and (b) conductivity in a correlated system. (i) Drude response (thin solid line). (ii) MIR band modeled by an overdamped oscillator (dashed line). (iii) Phonon described by the Lorentz model. Total conductivity $\sigma_1(\omega)$ and reflectivity $R(\omega)$ are shown by the thick solid line. From [30].

the Drude model or under the influence of correlations in the extended Drude description (Sec. 2.1.2). For organic conductors the response of the (renormalized) charge carriers in the conduction band typically spans up to about 6000-10000 cm^{-1}. The metallic response is given by the Drude contribution. Usually a plasma frequency in the order of $\omega_p/(2\pi c) \approx 100 - 1000$ cm^{-1} is found. But its value strongly depends on the electronic correlations and can be significantly reduced. Scattering rates $1/(2\pi c \tau)$ are in the range of 10 cm^{-1} to several 100 cm^{-1}. The deviations from a simple Drude peak are fingerprints of strong interacting charge carriers. Reviewing the extended Drude description gives additional information about the nature of the interactions (Sec. 2.1.2).

In addition to the Drude response the organic conductors show a broad contribution in the mid-IR. Commonly a huge band located around 1500-3000 cm^{-1} is observed. Phenomenologically it can be fitted with an over-damped Lorentz oscillator (Sec. 2.1.1.1). The origin of such bands can be different. It can be assigned to interband transitions in case a of a gap opening due to a dimerization of the molecular structure [31]. But also in non-dimerized structures such bands are observed, e.g. in the low temperature charge ordered θ-phases [32]. In that case they are assigned to the transition across the charge order gap. But also in non charge ordered phases at room temperature the band is present. More sophisticated models based on the (extended) Hubbard model, which are also used to explain the findings within this thesis (see Sec. 3), take into account the influence of on- and intersite Coulomb correlations, U and V, on the band structure [8, 33, 34]. There the density of states shows incoherent bands located at $\pm U/2$, the upper and lower Hubbard bands (also c.f. Fig. 2.13). Transitions between these bands give rise to a mid-IR charge transfer band. It describes the energy to overcome the Coulomb repulsion to hop on an occupied lattice site. Taking V into account gives rise to similar density of states effects. The details of these changes give the information about the dynamics of the electronic system. This picture, working on the Hubbard model, is well established in the half-filled systems and therefore shortly introduced in the subsequent section. It reveals the on-site repulsion U as driving force. The success in the description of these systems motivates to use the similar model in the quarter-filled systems. Therefore the extended Hubbard model, where the influence of the inter-site correlation V has to be taken into account, is used to describe the experimental results. Further contributions to the optical spectrum are given by lattice phonons of the crystal and vibrational modes of the molecules. Phonons are expected at very low frequencies below 200 cm^{-1} due to the heavy mass of the molecules occupying the lattice sites. The molecular vibrations are usually found in the 400-3000 cm^{-1} regime. They can be fit using the Lorentz oscillator (Sec. 2.1.1) or, depending on their coupling to the electronic background, with Fano lineshape (Sec. 3.3.2).

2.2. Low-dimensional organic molecular conductors as model system for correlated electron systems

The low-dimensional organic conductors and superconductors are a quite young class of materials which show a rich variety of interesting properties and so give an insight into the intriguing fields of physics like unconventional superconductivity, charge or spin density waves, charge or magnetically ordered states. In the focus here is the superconducting state and the interplay with the charge order process. The following chapter gives an overview of organic molecular crystals in general and especially their structures and electronic properties. Their main physical properties are due to strong electron correlations. The experimental evidences for that, especially in optics, are presented. A special focus is put on quarter-filled systems which are of the main interest of this thesis since they represent an ideal system to investigate the interplay of superconductivity and charge order phenomena due to electron-electron interaction.

2.2.1. Structure and electronic system

Most of the organic charge transfer salts are grown by electro crystallization from the solution resulting in high quality single crystals. The typical structure of these organic systems is presented here for the investigated BEDT-TTF based salts. The BEDT-TTF molecule, shown in Fig. 2.5, is the building block and acting as electron donor. The molecules have a large overlap of their π-orbitals to the neighboring molecules forming stacks. There is also an overlap to the neighboring molecule in the adjacent stacks; a two-dimensional conducting layer is formed. There is no direct orbital overlap from the BEDT-TTF layer to the anion layer (Fig. 2.6) which acts as electron acceptor, though there is a very weak Van-der Waals interaction. In that way the whole systems can be

2. Organic conductors as correlated electron systems

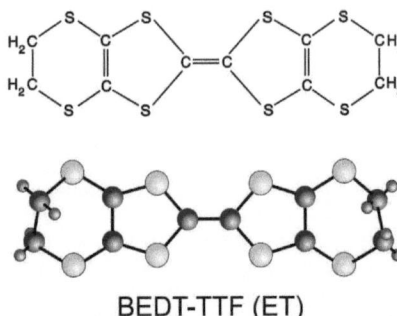

BEDT-TTF (ET)

Figure 2.5.: The bis(ethylenedithio)tetrathiafulvalene molecule abbreviated as BEDT-TTF or ET; the building block of the quasi two dimensional ET charge transfer salts. From [35].

regarded as a quasi-2D electron system. As pointed out in Sec. 2.1.2 this low dimensionality of two-dimensional conduction bands enhances the electron density and therefore the influence of electronic correlations.

The crystal structure of the main compound investigated in this thesis, the organic superconductor β''-(BEDT-TTF)$_2$SF$_5$CH$_2$CF$_2$SO$_3$, is shown in Fig. 2.7 (a). The prefix β'' stands for the specific pattern the molecules are oriented within the conducting plane (Fig. 2.7 (b)). An overview of different arrangements of molecules in the conducting layer is shown in Fig. 2.8. Depending on the arrangement within the unit cell the orbital overlap of the neighboring molecules differs. Therefore, in addition to the in-out of plane anisotropy of the system, an additional in-plane anisotropy could appear. The electronic band structure of the molecular solids can be viewed in a simplified model. The band formation is described similar to atomic crystals. In molecular crystals the atomic orbitals are replaced by the molecular orbitals. The overlap integrals t depends on the local amplitudes of the wave function of the molecular orbital and the spatial overlap to the neighboring orbital. For large molecules t is relatively small compared to dense packed atomic crystals of simple metals with large local amplitudes [37]. This satisfies a view of well separated bands; the energy separation of the molecular orbitals within one molecule is larger than the transfer integral between two molecules $\Delta > t$. Therefore it is sufficient to describe the molecular orbitals by the Highest Occupied Molecular Orbital (HOMO) and the Lowest Unoccupied Molecular Orbital (LUMO). The overlap between the molecules gives rise to the HOMO and LUMO band. The electronic properties of the organic conductors are dependent on the HOMO band. Its filling and geometrical structure determine the basic electronic band properties. Its structure is due to the anisotropic shape of the molecular orbitals, especially the π-orbital. The arrangement of the molecules within the unit cell leads to a different overlap between the molecular

2.2. Low-dim. organic conductors as model system for correlated electron systems

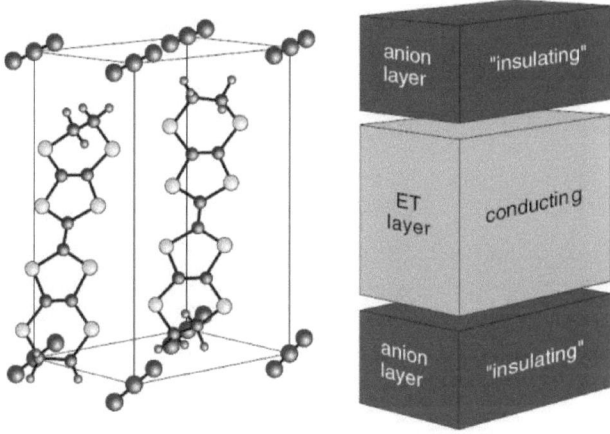

Figure 2.6.: The layered structure of the ET charge transfer salts. The quasi two dimensional conducting layers of ET molecules are separated by the insulating anion layers. From [35].

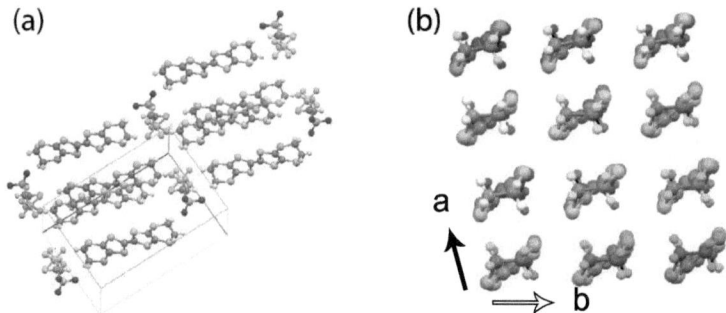

Figure 2.7.: (a) The molecular arrangement of β''-(BEDT-TTF)$_2$SF$_5$CH$_2$CF$_2$SO$_3$. (b) The β''-structure of the BEDT-TTF molecules in the conducting layer. Visualization performed with the Mercury software [36].

2. Organic conductors as correlated electron systems

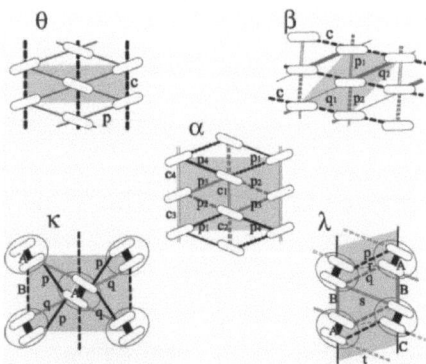

Figure 2.8.: Schematic view of the molecular arrangement within the conducting plane for different structures including their transfer integrals. Depending on the subfamily in the β- and α-phases some transfer integrals are the same. They can be mapped to an anisotropic triangular lattice according to Tab. 2.1. The dimers in the κ- and λ -phase can be seen as one lattice site. The β''−structure of the main crystals in this thesis is given in Fig. 2.7 which is the same as the β−system with different degree of dimerization. From [16].

sites and therefore to the electronic anisotropy found in molecular solids. That makes the structural arrangement a major point in the basic properties of these systems [16]. On the other hand the band filling is given by the amount of charge transfer and the stoichiometry. In (BEDT-TTF)$_2$X, which is a main representative of A$_2$B compounds, the anion X is fully charged as -1e. Therefore the BEDT-TTF molecules valence is $+1/2e$ in average. In the HOMO picture this results in a quarter-filled band of holes or 3/4-filled band of electrons as sketched in Fig. 2.9 (a). This situation is equivalent to the quarter filled case shown in Fig. 2.9 (b). In some systems there is a strong dimerization where two molecules overlap stronger to each other than to the next pair of molecules. In that case the pair of molecules, the dimers, are considered as one lattice site. The unit cell doubles and a dimerization gap opens leading to an effectively half-filled band (Fig. 2.9 (c)).

The different structures are shown in Fig. 2.8. The easiest one is the θ-phase with two molecules per unit cell and two different transfer integrals along the different directions. In the β-phases the transfer integrals are not uniform anymore and alternate along one direction leading to a band splitting. As subtypes β''-, β-, and β'-phases (in that order) have an increasing amount of anisotropy among their transfer integrals. In the case of κ- or λ-phases, one transfer integral is largest, leading to dimerization, where the two connected molecules (the dimer) can be regarded as one lattice site. Due to that dimer-

2.2. Low-dim. organic conductors as model system for correlated electron systems

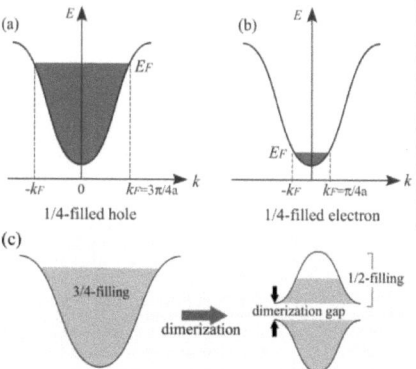

Figure 2.9.: Easy scheme for the band of A_2B compounds quarter filled with (a) holes or (b) electrons. (c) Dimerization of the quarter filled band due to the doubling of the unit cell leads to the opening of a dimerization gap and an effectively half filled band. From [16].

ization the unit cell doubles and a dimerization gap opens. Going from the κ- to the λ-phase is equivalent to going from the θ- to the β-phase just with the dimers instead of the single molecules on the sites. The increased anisotropy results in a band splitting. Due to dimerization these systems are effectively half filled.
Another type of structure is the α-phase; a four band system due to four molecules in the unit cell but without a dimerization gap. There are two subtypes, the α-I_3- and the α-MHg-type, where $M = (K, Rb, Tl, ...)$ stands for simple alkaline or alkaline earth metals. The difference is the degree of anisotropy and band splitting like between κ and λ-phases just without dimerization. The existence of different electronic ground states and how that is influenced by the underlying structure of the system is heavily investi-

	θ	β	κ	λ	α
t	t_c	t_c	$t_B/2$	$2(t_p + t_q - t_r)$	$t_{p1}/2, t_{p2}/2$
t'_1	t_p	t_{p1}	$(t_p - t_q)/2$	t_s	$(t_{p1} - t_{c1})/2$
t'_2	t_p	t_{p2}	$(t_p - t_q)/2$	t_t	$(t_{p3} - t_{c3})/2$
t'_3	t_p	t_{p3}	$(t_p - t_q)/2$	t_C	$(t_{p2} - t_{c2})/2$
t'_4	t_p	t_{p4}	$(t_p - t_q)/2$	t_B	$(t_{p4} - t_{c4})/2$

Table 2.1.: Mapping of the transfer integrals of the different structures (Fig. 2.8) to an anisotopic triangular lattice. The β''− mapping is the same as for the β−structure. From [16].

2. Organic conductors as correlated electron systems

gated, mainly based on the Hubbard or the extended Hubbard model [1, 8, 15, 16, 38–43]. Especially different types of charge order patterns (like stripes or checkerboard) can be explained based on the band structure (number of bands, band splitting, dimerization) and filling of the system. However the correlation dependent behavior in general is independent of the underlaying structure of the system. For a generalized description, all the different systems can be mapped to a triangular lattice according to Tab. 2.1. That highlights the electronic correlations as the essential parameter for the ground states among all the different structures. The structure itself controls the parameters of the effective correlations U/t and V/t.

2.2.2. Influence of electronic correlations

The structure and electronic band picture determine the basic properties of the organic conductors. But the influence of the effective electronic correlations is of major importance. They are the driving force of the dynamics in the low dimensional organic systems. Not all properties can be reduced to structural and band structure effects. As stated earlier, the two dimensionality already enhances the electron density and therefore electronic correlations play a pronounced role. Typical energy scales are as follows: Values for the transfer integrals are 10 meV to a few 100 meV based on x-ray structures with extended Hückel calculations [16, 44]. That is also in agreement with values from optical measurements of around 100 meV as e.g obtained in Ref. [45] via the plasma frequency. The effective Coulomb interaction values can be estimated to be $V \approx 0.1$ eV up to 0.5 eV and $U \approx 1$ eV. They are given by the Coulomb interaction between the electrons reduced by the effective on-site or inter-site screening of the molecular orbitals. The on-site Coulomb repulsion can be taken from the difference in the first and second oxidation potential resulting in the electron affinity which can be measured by photoelectron spectroscopy or cyclic voltammetry measurements [46, 47]. Further information about balanced U and V is taken from optical spectra [48] and theoretical calculations like Ref. [49] and references in there. Based on the two-dimensional square lattice, the bandwidth is calculated to be $W = 8t$ (or $W = 10t$ for the triangular lattice) within a tight binding model. With $U \gg V \geq W$ the typical conditions for a correlated electron system are given. Further, systems expected to be pure band metals show insulating behavior like charge order or even superconductivity.

The values of the on-site U is difficult to change. It is possible in a limited way by going from BEDT-TTF and TMTSF to BEDO, TMTTF, and TTF molecules. That changes the molecular orbitals and therefore the on-site screening. However, the physics of the systems is driven by the effective correlations U/t and V/t and therefore are also influenced by the structure via the overlap integral. Closely related to that is the change of the effective correlations by applying physical or so called chemical pressure. The physical pressure squeezes the crystal and enlarges the orbital overlap. That enhances the transfer integral t and reduces the effective correlation U/t, V/t. The chemical pressure is applied by changing the counter ion. A reduced or enhanced anion size alters the lattice constants and therefore the overlap and the effective correlations. The structure

2.2. Low-dim. organic conductors as model system for correlated electron systems

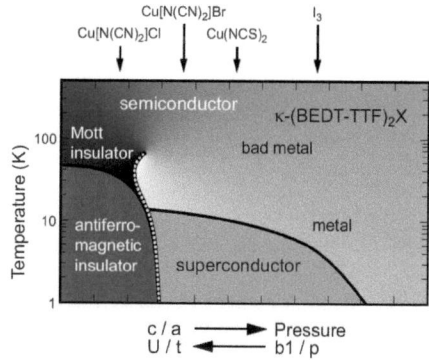

Figure 2.10.: Phase diagram of different κ-salts of BEDT-TTF. The correlations can be tuned by chemical pressure changing the overlap and therefore the transfer integrals. Instead the same effect can be reached by applying external pressure which alters the band overlap due to a change in the lattice parameters.

of the anion can also influence the arrangement of the molecules within the conducting plane which directly alters the orbital overlap, too.

The driving influence of the electronic correlations is evident in the half-filled κ-systems, e.g. These are among the most intensively studied compounds [8, 50–53]. A well established phase diagram (Fig. 2.10) is obtained from several studies on different compounds, changing chemical and physical pressure to tune the correlations. Reducing the effective correlations an antiferromagnetic insulator can become superconducting or metallic. Within a more unified view [16, 54] the influence of the on-site correlation to different structures, which determine band overlap and dimerization, tries to explain the different observed phases like paramagnetic metallic, charge ordered, and antiferromagnetic as overviewd in Ref. [15], e.g. The dimerization properties to a great extent also change the band filling from quarter- to effectively half-filled systems. Then the influence of the correlation is expected to display in a different way as sketched in Fig. 2.11. Among BEDT-TTF$_2$X systems the dimerized, half-filled systems tend to order para- or antiferromagnetic with increased correlation. The non-dimerized, quarter-filled systems tend to charge-order instead. One goal of the thesis is to establish and explain a phase diagram for quarter-filled systems similar to the one for half-filled systems.

2.2.3. Optical properties of half-filled systems

In their optical properties the half-filled κ-phase systems are characterized by a broad mid-IR band around 2500 cm^{-1}-3200 cm^{-1}. At high temperatures, no Drude response is found even though the systems are highly conductive within the plane. For exam-

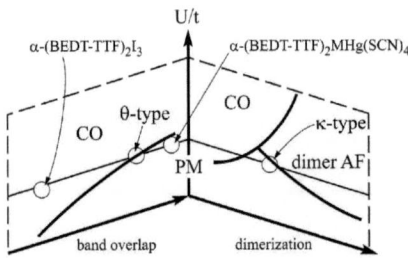

Figure 2.11.: Schematic phase diagram of a unified view of (BEDT-TTF)$_2$X for $V/U \approx 0.25$. It shows the trend of the organic BEDT-TTF$_2$X systems to charge order for increased correlations in the case of non-dimerized systems. With increased dimerization they tend to order magnetically. Either paramagnetic for a small band overlap or antiferromagnetic in cases of a large band overlap. From [15].

ple, in the superconductor κ-(BEDT-TTF)$_2$Cu(NCS)$_2$ only below 50 K a narrow Drude peaks shows up [55]. On cooling, the low-frequency spectral weight increases showing the increasing coherent transport. That is typical for a bad metal behavior found in different correlated systems [56]. Similar spectral effects are obtained for other κ-phases, e.g. the κ-(BEDT-TTF)$_2$Cu[N(CN)$_2$]Br$_x$Cl$_{1-x}$ series. Their temperature dependence of the optical conductivity is presented in Fig. 2.12. A good description of the κ-phases is based on the Hubbard model. Dynamical mean-field theory applied to a frustrated two-dimensional lattice at half filling with strong on-site correlations U is able to describe the conductivity spectrum [33]; the spectral contributions in the Mott insulating and metallic case can be understood based on the schematic view of the density of states given in Fig. 2.13 (a) and (b) respectively. They emerge from the results of the half-filled two-dimensional Hubbard model. For high U/t the system is insulating due to the presence of the Mott gap of $U - W$. It separates two bands, the lower and upper Hubbard band which are located at $-U/2$ and $U/2$. In the optical response this results in a band of width $2W$ located around U due to the transitions from the lower to the upper Hubbard band. The conducting case (Fig. 2.13 (b)) shows a quasi-particle peak in addition to the Hubbard bands. The additional spectral component around $U/2$ results from transitions from the lower Hubbard band to the quasi-particle peak and from the quasi-particle band to the upper Hubbard band. The presence of a Drude peak at low frequencies is due to the quasi-particle peak itself at the Fermi level. The peak and so the Drude response is suppressed on increasing U/t pushing the system towards the Mott transition where the quasi particle peak disappears and the system becomes insulating. A detailed study of the κ-(BEDT-TTF)$_2$Cu[N(CN)$_2$]Br$_x$Cl$_{1-x}$ series was performed [53, 59] to map out the phase transition from the antiferromagnetic insulator to the superconductor in the phase diagram Fig. 2.10. The spectral components based on a dimer model applied to the Hubbard model could be disentangled (Fig. 2.14) and followed by temperature

2.2. Low-dim. organic conductors as model system for correlated electron systems

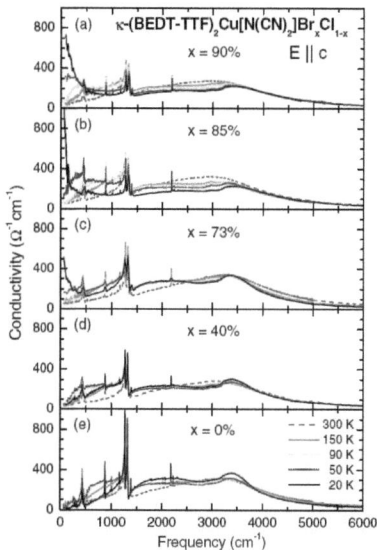

Figure 2.12.: Temperature dependent optical conductivity in the κ-(BEDT-TTF)$_2$Cu[N(CN)$_2$]Br$_x$Cl$_{1-x}$ series. From bottom to top the Br content x increases. The different chemical pressure reduces the effective onsite-correlations and the system shows a more and more metallic behavior at low temperatures. From [53].

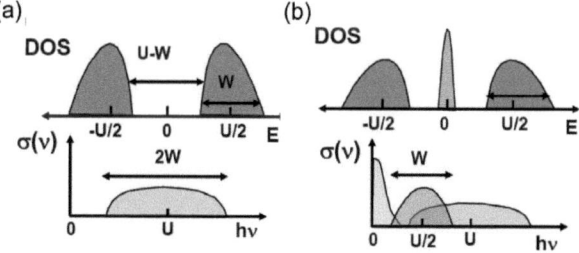

Figure 2.13.: Schematic density of states and optical conductivity for the Hubbard model in the (a) insulating and (b) conducting case. Due to the on-site Coulomb repulsion U the upper and lower Hubbard bands appear around $\pm U/2$. After Kotliar et al. [57].

2. Organic conductors as correlated electron systems

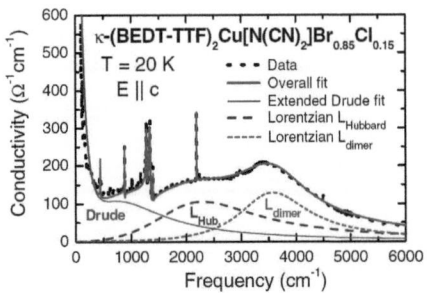

Figure 2.14.: Disentangling the optical contributions in κ-(BEDT-TTF)$_2$Cu[N()CN)$_2$]BR$_{0.85}$Cl$_{0.15}$. The mid-IR band shows contributions due to charge transfer in the dimers L_{dimer} and due to the transitions between the Hubbard bands due to strong electronic correlations $L_{Hubbard}$. From [53, 58].

(Fig. 2.12). One contribution originates from the charges localized on the BEDT-TTF dimers. This intradimer charge-transfer is coupled to the intramolecular vibrations. The second contribution is assigned to correlated charge carriers of the system close to the metal-to-insulator transition. They can be well described based on DMFT calculations ($U/t = 10$) [60]. The effective mass is found to diverge as the metal-insulator transition is approached. For parts of the phase diagram for the metallic species at low frequencies Fermi-liquid behavior could be observed and quasiparticles are gradually destructed with increasing temperatures. That is typical for a bad metal behavior. The successful disentangling and description of the correlated charge carriers in the half-filled organic conductors in this framework motivates to transfer these methods also to the quarter-filled systems.

2.2.4. Optical properties of quarter-filled systems

The quarter filling in the BEDT-TTF$_2$X series is directly due to the charge transfer of effectively half an electron from the BEDT-TTF molecule to the acceptor anion X. There is also a strong influence of the electronic correlations to the ground state of the system. But by just taking the effective on-site Coulomb repulsion U/t into account, which drives the dynamics of the half-filled systems, the quarter-filled systems should stay metallic for all values of U/t. However for large values of U and a certain strength of the nearest-neighbor interaction V insulating charge ordered states can appear.

2.2. Low-dim. organic conductors as model system for correlated electron systems

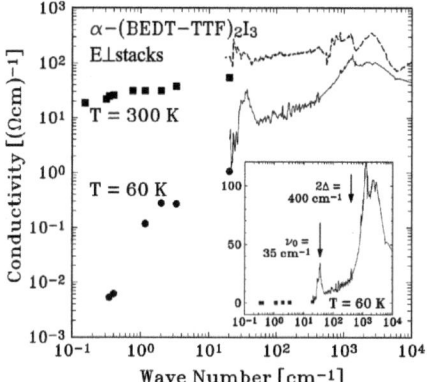

Figure 2.15.: Optical conductivity of α-(BEDT-TTF)$_2$I$_3$ perpendicular to the stacks above (300 K) and below (60 K) the metal insulator transition at 135 K. Below the transition an energy gap at 400 cm^{-1} appears as enhanced in the inset. From [61].

Charge ordered insulators

First experimental evidence for such an charge ordered state was found in ESR measurements of θ-(BEDT-TTF)$_2$CsZn(SCN)$_4$ [62] and NMR experiments on several θ-phases [63] and α-(BEDT-TTF)$_2$I$_3$ [64]. In optics the Drude spectral weight is found to decrease as the charge order transition is approached and the electronic spectra become insulating or semiconducting [32, 65–67]. In addition, in the metallic regime optical features in the low frequency region appear which triggered a lot of theoretical investigations [15, 17–19, 44, 54, 68, 69]. They result in a picture of an intersite Coulomb repulsion V driven phase transition as discussed in more detail in Sec. 3.

Among the quarter-filled systems, α-(BEDT-TTF)$_2$I$_3$, which is also under investigation within this thesis, is the most prominent example for a metal-insulator transition ($T_{MI} = 135$ K) due to charge order. Optical studies [61] reveal a drop in the conductivity of orders of magnitude below the transition as displayed in Fig. 2.15. An optical gap opens at $2\Delta = 400$ cm^{-1} indicating charge localization. The mode appearing at 35 cm^{-1} is assigned to a lattice phonon of the molecular crystal. No coupling of this mode to the charge order is seen or reported [71]. This is different for the vibrational modes of the BEDT-TTF molecule itself. IR absorption measurements of α-(BEDT-TTF)$_2$I$_3$ powder reveal a splitting in the frequency dependent C=C double bond mode $B_{1u}(\nu_{27})$ of the BEDT-TTF (Fig.2.16) [70] right at the charge order transition. This is directly related to the charge disproportionation between the molecular sites. The theoretical background of that is discussed in Sec. 3.3.1. This charge disproportionation is also confirmed in

2. Organic conductors as correlated electron systems

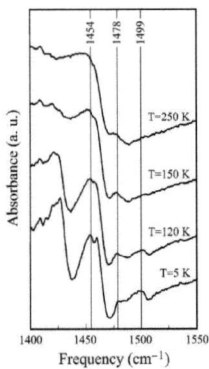

Figure 2.16.: Optical absorption of α-(BEDT-TTF)$_2$I$_3$ powder. Splitting of the $B_{1u}(\nu_{27})$ vibrational mode at the metal insulator transition at 135 K. Above the phase transition only a single mode at 1478 cm^{-1} is reported while below the transition the mode splits due to charge-order in the system with contributions at 1449 and 1457 cm^{-1}. From [70].

Raman scattering experiments [72, 73], and partly in NMR spectra [64] which also gives hints for fluctuating charge order preceding the transition. The details of this charge order transition will be presented in more detail in the materials section. The experimental data so far is in agreement with a formation of a horizontal stripe charge order pattern. The related charge redistribution that takes place on crossing the transition is subject to investigations of this thesis.

Charge order also takes place in the θ-(BEDT-TTF)$_2$MM'(SCN)$_4$ family. Here a modifications of the dihedral angel, defined in Fig. 3.1, between the molecules in adjacent stacks changes the overlap integrals due to different combination of metal ions (M=Rb, Tl, Cs and M'=Zn, Co). At room temperature they behave metallic. On cooling they undergo a metal to insulator transition at different temperatures. The θ-(BEDT-TTF)$_2$RbZn(SCN)$_4$ salts show a sharp transition at $T_{MI} \approx 190$ K accompanied by a structural phase transition; a slight dimerization of the stacks takes place. In optics (Fig. 2.17), although no distinct Drude peak is seen, a transfer of low frequency spectral weight to higher frequencies at the transition is reported in Ref. [32]. However Ref. [65] reports an optical gap that opens at 300 cm^{-1} for θ-(BEDT-TTF)$_2$RbZn(SCN)$_4$ as seen in Fig. 2.18. Like in α-(BEDT-TTF)$_2$I$_3$ a horizontal stripe charge order pattern (Fig. 4.28) is expected to appear [32, 65] and is discussed in combined theoretical and experimental work on the vibrational spectra [74, 75] or NMR studies [63, 76, 77] of this family. In these measurements also precursors of charge fluctuations are found. The details of this phase transition are also presented in the materials section. Especially

2.2. Low-dim. organic conductors as model system for correlated electron systems

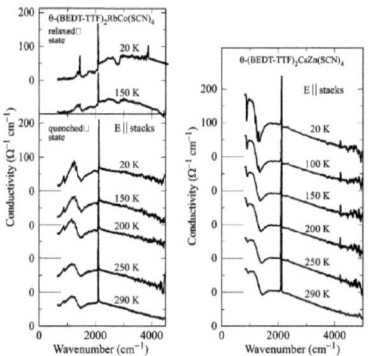

Figure 2.17.: Optical conductivity of θ-(BEDT-TTF)$_2$RbCo(SCN)$_4$ and θ-(BEDT-TTF)$_2$CsZn(SCN)$_4$. No distinct Drude peak is seen but low frequency spectral weight gets transfered to higher frequencies in the RbCo compound below 190 K while no changes are reported for CsZn. From [32].

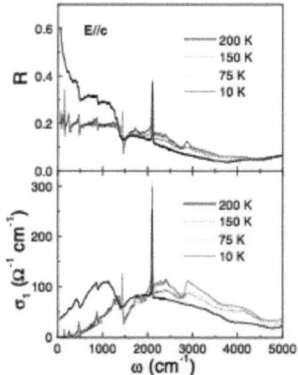

Figure 2.18.: Reflectivity and optical conductivity of θ-(BEDT-TTF)$_2$RbZn(SCN)$_4$. Below 190 K the reflectivity drops from a metallic like behavior to an insulating one. The optical conductivity starts drop below 2000 cm^{-1} and opens to a charge order gap at 300 cm^{-1}. From [65].

2. Organic conductors as correlated electron systems

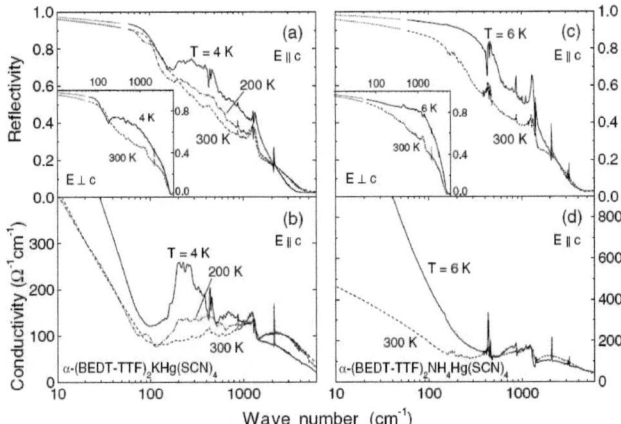

Figure 2.19.: Reflectivity and optical conductivity of (a-b) α-(BEDT-TTF)$_2$KHg(SCN)$_4$ and (c-d) α-(BEDT-TTF)$_2$NH$_4$Hg(SCN)$_4$. Below 200 K a reflectivity dip in the K-salt below 500 cm^{-1} leads to the formation of a conductivity feature around 200 cm^{-1} within a pseudo gap. At lowest frequencies a strong Drude peak is found. For the related NH$_4$-compound no pseudo-gap was observed. Just a large Drude response is present. From [67].

the comparison of θ-(BEDT-TTF)$_2$RbZn(SCN)$_4$ and α-(BEDT-TTF)$_2$I$_3$ is investigated within this thesis using vibrational IR spectroscopy.

Renormalized metals close to charge order

As already seen above, in the metal-insulator compounds, charge fluctuations preceding the transition might be of importance. Therefore, especially the metallic phase close to the charge-order transition is of interest. The family of α-(BEDT-TTF)$_2$MHg(SCN)$_4$ salts (M=K, Rb, Tl, and NH$_4$) are candidates located close to the charge order transition [79]. All are metallic with indications for density wave states at 8-12 K (which are still under discussion). The NH$_4$ compound even becomes superconducting at T_c =1 K [80, 81]. Fig. 2.19 shows the reflectivity and optical conductivity of the K and the superconducting NH$_4$ salt for different temperatures. Above 500 cm^{-1} no significant differences between the compounds are found. Below that frequency a reflectivity dip appears in the K-salt for temperatures below 200 K. That leads to a conductivity feature around 200 cm^{-1} within a pseudogap. At lowest frequencies a Drude peak of small spectral weight remains. In the superconducting NH$_4$ salt it is difficult to distinguish from the data if a

2.2. Low-dim. organic conductors as model system for correlated electron systems

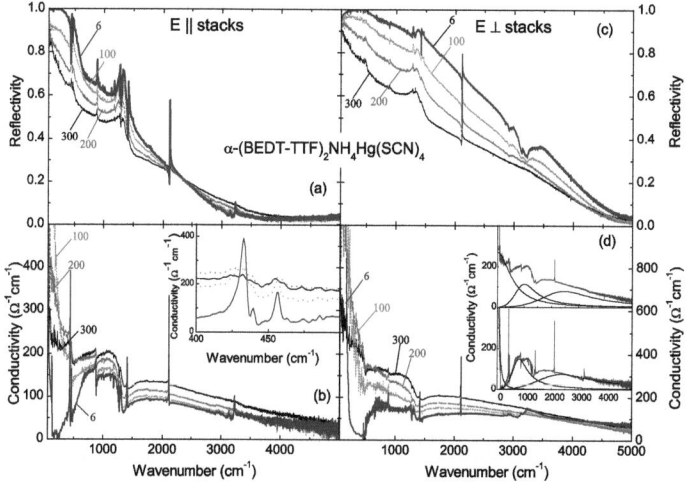

Figure 2.20.: Reflectivity and optical conductivity of α-(BEDT-TTF)$_2$NH$_4$Hg(SCN)$_4$. In contrast to the previous study Fig.2.19 at temperatures below 6 K a pseudo gap was found with a small Drude contribution below the gap. The inset in (d) shows the 3 contributions to the optical conductivity. From [78].

2. Organic conductors as correlated electron systems

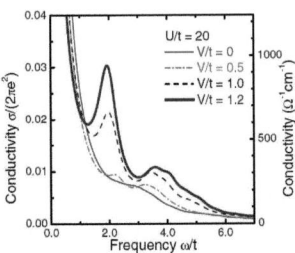

Figure 2.21.: The evolution of the optical conductivity as function of effective inter-site Coulomb repulsion V/t at the metallic side of the phase transition. The critical value for a charge order phase transition is $(V/t)_c = 2.2t$. Calculation as for the model described in Sec. 3. From [67].

pseudogap is absent in favor to a Drude peak of large spectral weight (Fig. 2.19) or if the pseudogap is present with a Drude response below (Fig. 2.20). In any case more spectral weight is found at low frequencies. The inset (d) in Fig. 2.20 shows the contibutions to the optical conductivity. Besides the Drude contribution and the large MIR band, a peak at 700 cm^{-1} is observed leading to the opening of the pseudo gap. In the following that band can be assigned as a charge fluctuation feature.

For a successful description of the spectra, a model based on nearest-neighbor interaction V driven charge fluctuations can be used as described in Sec. 3. The main features include a Drude response to describe the quasi free charge carriers and a MIR band to account for the localized charges due to the electronic correlations (Hubbard band). In addition a charge fluctuation feature at low but finite frequencies results from the interaction between the free and localized charge carriers. That is mainly due to the inter-site Coulomb correlations in the quarter-filled systems. Within this model optical conductivities are calculated as in Fig. 2.21. The comparison between the K- and NH$_4$-salt perfectly fits these calculated spectra assuming the superconducting NH$_4$-compound more on the metallic side. The pronounced localization and gap features of the K-compound shows the higher correlations present in the system $(V/t)_K > (V/t)_{NH_4}$. That is also in agreement with pressure dependent studies where more metallic states are favored on pressure because of the higher transfer integrals reducing the effective correlations. Indeed the K-compound becomes superconducting under high enough pressure [82]. Further the superconducting NH$_4$-salt is compared to the Tl- and Rb- compounds in Fig. 2.22 showing reflectivity and conductivity for various temperatures. All compounds show a Drude response already at room temperature. At the same time localization features are present. The charge fluctuation feature is located around 700 cm^{-1} and an overall grow of spectral weight is found on cooling. The mid-IR band looses spectral weight in favor to low frequencies. The redistribution of spectral weight between the interaction feature and the Drude response differs between the compounds:

2.2. Low-dim. organic conductors as model system for correlated electron systems

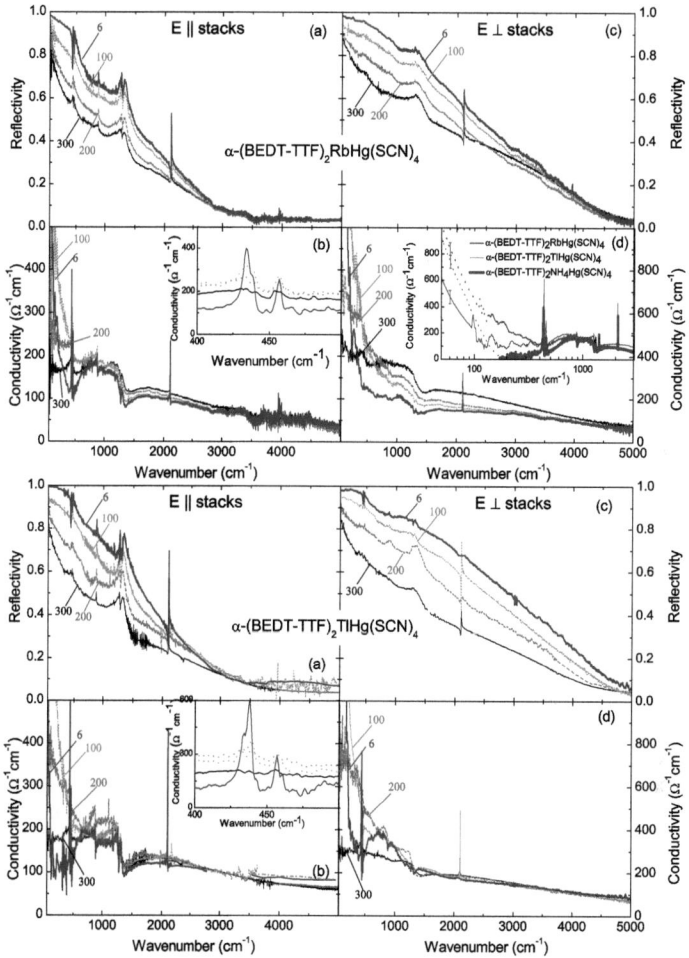

Figure 2.22.: Reflectivity and optical conductivity of (top) α-(BEDT-TTF)$_2$RbHg(SCN)$_4$ and (bottom) α-(BEDT-TTF)$_2$TlHg(SCN)$_4$. As in the related NH$_4$ compound Fig. 2.20 contributions due to Drude and the charge fluctuation band are present. In all cases the Drude contributions stays dominant. But in comparison to the NH$_4$ salt the the spectral weight on cooling in the Rb salt enhances more the charge fluctuation feature. On cooling the Tl salt first the spectral weight goes to both the Drude and the charge fluctuation feature while the latter is enhanced mainly at lowest temperatures. From [78].

2. Organic conductors as correlated electron systems

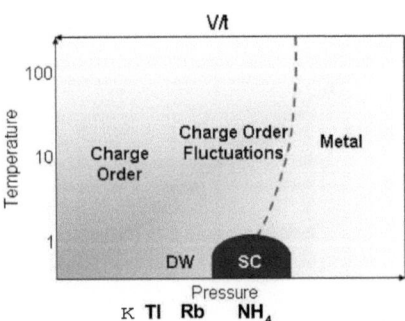

Figure 2.23.: Phase diagram for the members of the α-(BEDT-TTF)$_2$MHg(SCN)$_4$ family based on optical results. From [78].

In the superconducting NH$_4$-salt the metallic properties and interaction feature are of comparable spectral weight. On cooling, the metallic part starts to enhance more, in agreement of a theoretical prediction of a reentrant-like behavior (c.f. Sec. 3).
Compared to that, in the Rb-salt a stronger enhancement of the interaction feature is found with decreasing temperature.
For the Tl-compound the spectral weight transfer goes in favor to both the Drude response and the interaction feature. At lowest temperatures the strongest increase is in the interaction feature. But already at high temperatures a splitting of the vibrational charge sensitive modes of the BEDT-TTF molecules in Raman spectra points out the presence of an ordered state.
Based on the increase of the interaction features the salts are placed within a phase diagram as shown in Fig. 2.23. The K-salt is suggested at even higher correlations based on pressure dependence in [78] and the optical results in Fig. 2.19.
In a calculated phase diagram (Fig. 3.3), based on a nearest neighbor interaction driven charge-order model (c.f. Sec. 3), the strong enhancement of the interaction feature should lead to superconductivity. However, here a competing charge-order state is formed. This is caused by structural effects of the more complicated α-structure which might explain the density wave state (which origin and character is still under discussion) at lowest temperatures for the non-superconducting systems.

The (BEDT-TTF)$_2$SF$_5$RSO$_3$ series of renormalized and charge ordered metals

Another important series of quarter filled systems are the β''- and β'-systems of the (BEDT-TTF)$_2$SF$_5$RSO$_3$ family, where R stands for an organic rest. The superconducting member β''-(BEDT-TTF)$_2$SF$_5$CH$_2$CF$_2$SO$_3$ is in the focus of this thesis. Other main compounds are the isostructural metallic β''-(BEDT-TTF)$_2$SF$_5$CHFSO$_3$, which also is

2.2. Low-dim. organic conductors as model system for correlated electron systems

anion R	structure	dc behavior	anion pocket	phase transition
CH_2CF_2	β''	metal - sc T_c=5.4 K	ordered	no
CHF	β''	metal	ordered	no
$CHFCF_2$	β''	metal - insulator T_{MI}=190 K	disordered	anion order
CF_2,CH_2	β'	insulating	disordered	dimerization

Table 2.2.: Overview of the different compounds of the (BEDT-TTF)$_2$SF$_5 R$SO$_3$ series.

investigated here, and the β''-(BEDT-TTF)$_2$SF$_5$CHFCF$_2$SO$_3$, which undergoes a metal insulator transition around 190 K. Further there are insulating CH$_2$ and CF$_2$ members, which show strong dimerization and therefore are already different from the structural point of view having a β'-structure. An overview is given in Tab. 2.2. Details about experimental findings are found in the materials Section 4.1. Compared to the before presented α-(BEDT-TTF)$_2$MHg(SCN)$_4$-family neither in the metallic nor superconducting compound a Drude response has been found so far [66, 83, 84]. This is assigned there due to localization effects of the electronic system, despite the superconducting transition. It can be also argued to be due to the proximity to the charge order transition based on the high correlations. It would reduce the Drude behavior and leave only a small contribution at very low frequencies. That is interesting in particular for an interplay between charge order and superconductivity in comparison to the more competing picture seen in the α-systems. Its fairly simple structure and the fact that in the main compound, the superconductor β''-(BEDT-TTF)$_2$SF$_5$CH$_2$CF$_2$SO$_3$ and its isostructural metal β''-(BEDT-TTF)$_2$SF$_5$CHFSO$_3$ no structural changes take place, allows to focus on the electronic properties leaving out structural effects in first approximation.

2.2.5. One fifth-filled systems

As already stated above and seen in the generalized view of the sketch in Fig. 2.11 the resulting ground states are different for different band-filling. That is because the filling determines the position of the Fermi level within the band structure and therefore directly alters the physical properties. That is why quarter-filled systems show a different physics than the half-filled systems introduced before. It would be interesting to investigate the crossover between these regimes. But especially gradual changes to explore the influence to the ground states are rare. An external control via field-effects was tried but still remains a major challenge [85–87] and results are still difficult to judge whether effects are really field-effects, altered correlations, or just simply heating.

However, it is possible to change the filling away from the quarter-filled case by doping

2. Organic conductors as correlated electron systems

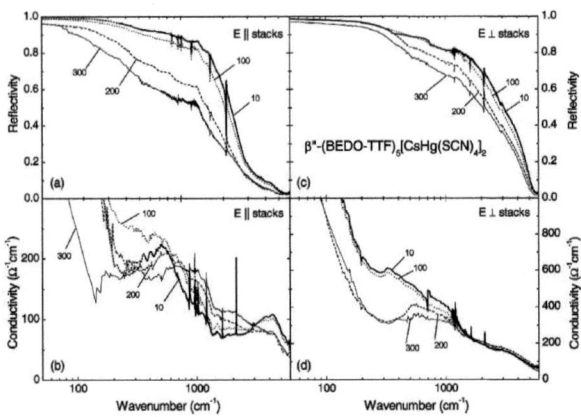

Figure 2.24.: Optical response of the fifth-filled β''-(BEDO-TTF)$_5$[CsHg(SCN)$_4$]$_2$. Already at room temperature a pronounced Drude peak is present and remains dominant at all temperatures. From [88].

the charge order pattern with charge carriers. This leads to an increase in the Drude response. The optical response of the fifth-filled β''-(BEDO-TTF)$_5$[CsHg(SCN)$_4$]$_2$ is shown in Fig. 2.24 [88]. Besides the Drude peak (present for all temperatures) there is a feature around 700 cm^{-1}. It is assigned to the charge fluctuation band due to the remaining electron-electron interactions. In contrast to quarter-filling, no metal-insulator transition is expected. The doping always allows mobile carriers to hop to a non-occupied site. Therefore a finite Drude response for all correlation values and at all temperatures is expected. In an interplay picture of charge-order and superconductivity the interaction features should be also present. They are just smaller in spectral weight and dominated by free carrier response. Assuming the 700 cm^{-1} peak which softens to 350 cm^{-1} on cooling to be the interaction feature, these results are in agreement with the picture of charge-order fluctuation mediated superconductivity which is proposed.

The temperature dependence of the Drude spectral weight in half-, quarter-, and fifth-filled systems close to the Mott transition is compared in Ref. [89]. Fig. 2.25 shows that quasi-particles in the half-filled compound are observed only well below 100 K but with a strong enhancement of the Drude spectral weight on cooling. That is typical for a bad metal. The quarter-filled systems show a Drude-like peak already at room temperature with only slight increase on decreasing temperature. That is in agreement with the proposed charge fluctuation picture. For the fifth-filled systems the Drude contribution is nearly temperature independent. Doping free charge carriers do not allow to form commensurate charge order patterns and the Drude spectral weight stays conserved.

2.2. Low-dim. organic conductors as model system for correlated electron systems

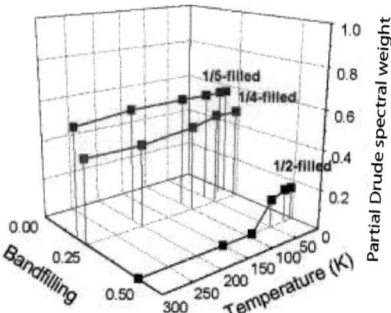

Figure 2.25.: Drude contribution for half-, quarter-, and fifth-filled systems as evolution with temperature. From [89].

3. Interplay of charge-order and superconductivity

The experimental findings in the class of the quarter-filled systems indicate a rich phase diagram. The extended Hubbard model on a square lattice turns out to be a simple model that is powerful enough to explain the main features of this phase diagram. It can reproduce the metal-insulator transition driven by the nearest-neighbor Coulomb interaction V. The model is able to describe superconductivity, arising from quasi-particle scattering at charge fluctuations. Ross H. McKenzie, J. Merino, A. Greco, and M. Calandra *et al.* propose an interplay of charge order and superconductivity to explain the experimentally observed electron dynamics [17–19, 44, 69].

An overview is given how these metallic, charge ordering, and superconducting effects are seen in optics. In addition, to probe the charge order, IR and Raman vibrational spectroscopy are used to follow the charge sensitive molecular vibrations as function of temperature. This unique possibility for organic molecular crystals is explained and discussed.

3.1. Description of 2D quarter-filled systems using the extended Hubbard model

As discussed in Sec. 2.2, for organics in general, and in Sec. 4 for the investigated systems in particular, different ground states due to structural and especially correlation effects are present among the quarter-filled two-dimensional organic conductors. Experimental findings point towards a strong influence by charge order and charge fluctuations. McKenzie *et al.* [44] propose that the peculiar phase diagram of metallic, charge order, and superconductivity could arise due to the proximity to a charge ordering quantum critical point. To describe this physics and explain the basic experimental observations with a model as simple as possible, it turns out to be important to take the influence of the nearest-neighbor Coulomb interaction V into account. Starting from a Hubbard model $H = T + U$, in which the kinetic part of the electrons is described by T and the Coulomb interaction of the electrons on the same site by U, this leads to the extended Hubbard model $H = T + U + V$ as described in Sec. 2.1.3.

The two dimensional lattice as it is realized as triangular lattice in the most general case is shown in Fig 3.1. Other lattice types can be mapped to this (Tab. 2.1, Fig. 2.8). The overlap of the molecules determines the transfer integrals t to the neighboring sites. This also defines the energy scale. All correlation energies and frequencies ranges in the

3. Interplay of charge-order and superconductivity

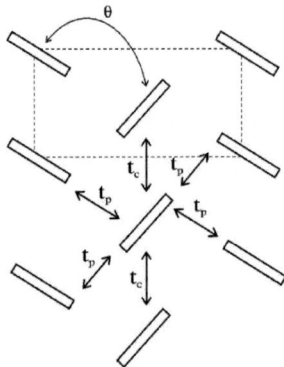

Figure 3.1.: Scheme of the molecular arrangement within the conducting plane for the θ-phase. The triangular lattice can also be represented as square lattice with hopping along one of the diagonals. From [44].

calculations are measured in units of t. In the simplest case the transfer integrals to the next neighbors are much bigger than to the next nearest neighbors $t_p \gg t_c$. Hence neglecting the diagonal hopping t_c reduces the triangular lattice to a simple square lattice with a transfer integrals to the nearest neighbors $t=t_p$. That is exactly the case for the θ-phases. The Hamiltonian for such a system is written like Eqn. (2.11) as

$$H = t \sum_{\langle ij \rangle, \sigma} \left(c^\dagger_{i\sigma} c_{j\sigma} + c^\dagger_{j\sigma} c_{i\sigma} \right) + U \sum_i n_{i\uparrow} n_{i\downarrow} + V \sum_{\langle ij \rangle} n_i n_j - \mu \sum_{i\sigma} n_{i\sigma} \qquad (3.1)$$

where the operator $c^\dagger_{i\sigma}$ describes the electron with spin σ at site i, n_i is the number operator, and $\langle ij \rangle$ sums over nearest neighbors. In addition to Eqn. (2.11) μ is the chemical potential introduced to control the band filling. In the case of organic charge transfer salts it was shown that the on-site as well as the nearest neighbor interactions are high and dominate the electron dynamics. That is why they are considered as strongly correlated electron systems. Since the on-site Coulomb repulsion U in the organics is the very dominant term it is unlikely that the molecular sites are double occupied. Therefore the system described with the Hamiltonian (2.11) can be reduced to a $t-V$ model which in leading order of $1/N$ expansion (see Appendix 14.1) can be written as

$$H = H^f + H^b + H^{f-b}, \qquad (3.2)$$

which splits into a fermionic, bosonic, and interaction term [17, 18, 44]. They represent the mean field solution, the fluctuations about the mean field solution and the coupling between these. The single contributions are summarized and discussed in the following.

3.1. Description of 2D quarter-filled systems using the extended Hubbard model

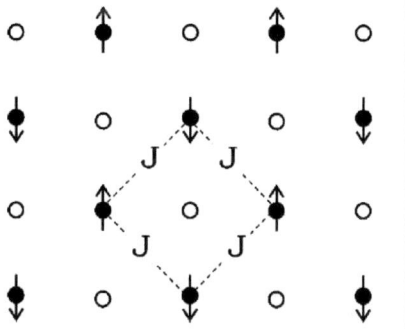

Figure 3.2.: Checker board or diagonal stripe like charge order pattern on the two dimensional square lattice established for correlations above the critical value $(V/t)_c$. The charge order leads to an exchange interaction J on a square lattice tilted by 45 degrees with respect to the original crystal lattice. From [44].

3.1.1. Nearest neighbor interaction driven charge order

The mean-field solution of the electronic system is represented by the fermionic part H^f of the Hamiltonian Eqn. (3.2). The electronic system is described by the actual charge of the interacting carriers and their distribution on the lattice. It results in a renormalized Fermi liquid where for quarter-filled systems the bandwidth is expected to be reduced to the half and therefore the effective mass enhanced by $m^*/m = 2$ [17, 44].

The second term of the Hamiltonian Eqn. (3.2), the bosonic part H^b, describes the influence of dynamical and spacial fluctuations around the mean-field solution (see Appendix 14.1 for details). They are present as fluctuations in the charge density. These fluctuations are driven by the effective nearest neighbor interaction V/t [17, 44]. They destabilize the Fermi liquid and, exceeding a critical value $(V/t)_c = 0.69$, a quantum phase transition occurs and turns the system into an insulating checker-board charge order state (Fig. 3.2). No Fermi surface nesting takes place: The charge order transition is a pure correlation effect due to the nearest neighbor interaction V. The results also hold for a triangular lattice with slightly different critical values of effective correlations [44]. That makes the model also applicable for the β'' which are in the focus of the experimental studies.

A detailed study of the metallic phase next to the charge order transition reveals a so called *charge-ordered metallic state* which is realized for $V_c^{CO} < V < V_c^{MI}$ [69]. Already prior to the metal insulator transition the formation of charge order patterns sets in but the system still shows a metallic response. The higher the correlations, the more charge order is established. Based on the calculations, the metal-insulator transition is expected to occur at nearly complete charge order of more than 75 %.

3.1.2. Superconductivity mediated by charge fluctuation

Following the idea that the system is close to a charge-ordering critical point, the question arises if there is the possibility for an instability toward a superconducting state. In fact, using the slave boson theory (Appendix 14.1 and Refs. [17, 18]), J. Merino and R.H. McKenzie pointed out that superconductivity can occur mediated via charge fluctuations driven by nearest neighbor interactions V [17]. In the effective Hamiltonian Eqn. (3.2), $H = H^f + H^b + H^{f-b}$, the last term describes the coupling of the renormalized Fermi liquid and the bosonic part which represents the fluctuations above the mean field solution due to the electron-electron repulsion. This interaction between the quasiparticles can be attractive and lead to Cooper pairing. The charge fluctuations (bosons) act here in analogy to phonons in the pairing of Cooper pairs in conventional superconductors. The renormalized Fermi liquid shows an instability to superconductivity for $V/t > 0.4$ at $T = 0$ and, in this case, the quasi particle interaction via charge fluctuations leads to an attractive potential that overcomes the Coulomb repulsion. According to their theory, an attractive interaction with d_{xy} symmetry can result in the formation of Cooper-pairs. As the correlations become stronger, above the critical value $(V/t)_c \approx 0.69$, the system turns into the charge ordered state, again. As for the charge-order transition, the model also holds for a triclinic lattice when the diagonal terms are included. That means it can describe β''−systems.

As a result, the calculated phase diagram (Fig 3.3) for the quarter-filled square lattice can explain the presence of a metallic, charge ordering, and superconducting state as a function of temperature T and effective nearest neighbor interaction V/t. As expected, without correlations, $V = 0$, the system is purely metallic. As soon as some interaction is turned on, charge fluctuations are present and lead to a renormalized Fermi liquid. It stays metallic until the critical effective correlation value for the charge ordering metal-insulator transition is reached. As described above, for effective correlations larger than $V/t = 0.4$ at $T = 0$ the charge fluctuations act as attractive interaction and turn the system superconducting. Further increase of the correlations increases the fluctuations and above the critical value $(V/t)_c = 0.69$ turn the system into the insulating charge order state. The superconducting state can also be destroyed by temperature. At higher temperatures stronger interactions and therefore higher effective correlations are needed. Interesting to notice is the back bending of the line describing the charge order transition: Starting at high temperatures and decreasing the temperature the critical value of correlations also decreases. That means less charge fluctuations are needed to destabilize the renormalized Fermi liquid. This is plausible since less thermal fluctuations have to be overcome. But going further down in temperature the calculations show a non-trivial reentrant behavior. The transition curve bends back and higher correlations are needed to enter the charge ordered state. In other words the systems turns more metallic, again.

Besides the pure charge fluctuation effects, which the above model is based on, there is the presence of an antiferromagnetic exchange interaction within the charge ordered phase [44] (see also Fig. 14.1). A. Greco et al. described the superconductivity based on a $t - J - V$-model taking into account the spin-spin interactions which are remanent

3.2. Dynamics and spectral traces of metal, charge order, and superconductivity

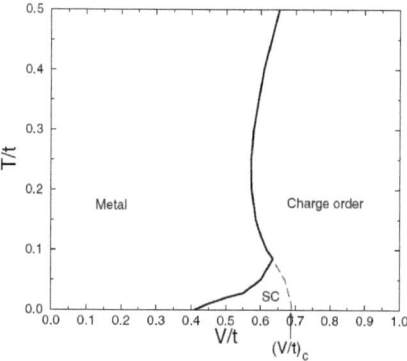

Figure 3.3.: Phase diagram for the extended Hubbard model calculated on a quarter-filled square lattice in the $U \to \infty$ limit. It shows the presence of a metallic, charge ordered, and superconducting phase. At low temperatures a back bending of the charge order transition line is found. From [17].

in the metallic phase close to the charge order [19]. It is shown that an increasing J leads to a decrease of the critical value V_c for the charge order transition. Further a large J also favors and stabilizes the d_{xy}-symmetry of superconductivity. In that sense, it can help to explain, even if not completely, the rather high T_c of several Kelvin found experimentally in the organic conductors. They cannot be explained just from the charge-charge interactions alone.

3.2. Dynamics and spectral traces of metal, charge order, and superconductivity

The intriguing interplay of the metallic, charge ordered, and superconducting phases revealed in the previous section is purely based on the electronic part of the system and its dynamics. To detect this dynamics, optical techniques are an ideal probe. They directly couple to the electrons via the dipole moment. Measuring the reflectivity over a broad frequency range, we can evaluate the optical properties from the optical conductivity as response function. But also the vibrational properties give rich information; in particular it is a perfect probe to investigate the charge disproportionation within the system. That is discussed in the next section. In the following the typical response of the electronic system expected for metallic, charge ordered and superconducting phases are discussed. Especially the metallic state close to the metal insulator transition, the so-called charge-ordered metal, is described in detail. It shows the main features of the interplay that is connected to superconductivity.

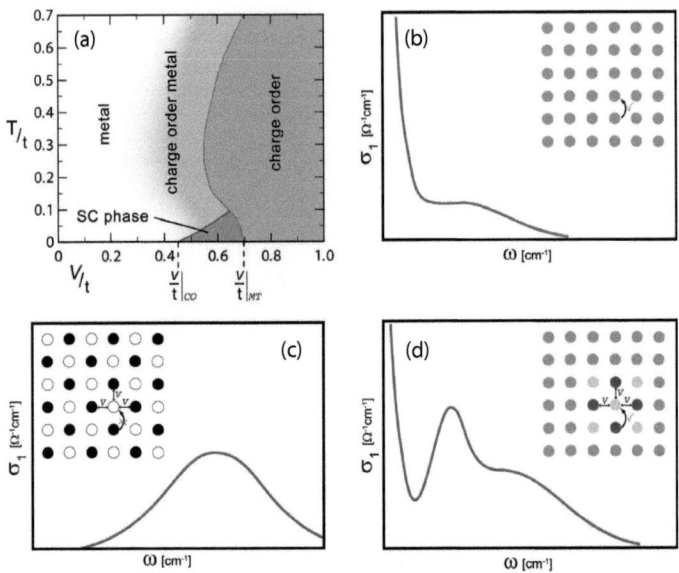

Figure 3.4.: Phase diagram (a) for the extended Hubbard model on a quarter-filled square lattice and the dynamical properties of the (b) metallic, (c) insulation charge ordered, and (d) charge ordered metal state. Besides the expected optical conductivities the electron distribution on the lattice is sketched as an inset for each case. After [17].

3.2.1. Metallic system

Neglecting correlation effects ($V = 0$) we are in the metallic phase of the phase diagram (Fig. 3.4 (a)). In a quarter-filled system that holds even for $U \neq 0$. On the lattice the charge carriers can move freely and are distributed in average with half a charge per site as sketched in inset of Fig. 3.4 (b). In the optical response a Drude peak is expected. In addition the presence of a charge transfer band in the MIR (assuming integrals of $t \sim 0.1$ eV typical for organic molecular systems) is due to the on-site Coulomb repulsion $U \sim 20t$ (Fig. 3.4 (b)). With increasing V the corresponding charge transfer band is expected to be enhanced (c.f. Sec. 3.2.3 about dynamical properties of the charge ordered metal).

3.2. Dynamics and spectral traces of metal, charge order, and superconductivity

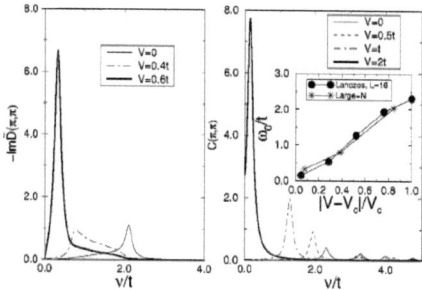

Figure 3.5.: Collective mode at critical wave vector (π, π) calculated with large-N method (left) and Lanczos diagonalization (right) at different correlations V/t. The inset shows the softening as the charge order transition is approached at V_c. From [18].

3.2.2. Charge order

In the other extreme of very strong correlations, the system is in a charge-ordered insulating state. On the lattice a checker-board like charge-order pattern is established. No free hopping is possible because of the large intersite Coulomb repulsion V from the neighboring sites (inset Fig. 3.4 (c)). Therefore no Drude peak is expected. Instead a charge-order gap opens below the enhanced charge transfer band (Fig. 3.4 (c)). This band is located around $\omega = 3V$, which is the energy needed to overcome the Coulomb repulsion for hopping to a free site. In Fig. 3.7 this situation is calculated for $V = 2.3t$ (where the critical value for the charge order transition is $V_c = 2t$) and $U = 20t$. The enhanced mid-IR band is seen at frequencies of $\omega = 3V \approx 6t$ with the charge-order gap below it.

3.2.3. Dynamical properties of the charge ordered metal

Most interesting and important for the formation of the superconducting ground state is the dynamics of the metallic state close to the metal-insulator transition; the so-called charge-ordered metallic state. A collective charge mode, shown in Fig. 3.5, represents the charge response of the fluctuations towards the checker-board charge-order. This excitation is not direct accessible as pure electronic mode in optics due to its (π, π) wavevector in k-space. An additional excitation providing the (π, π) wavevector would be needed. The wavevector of the collective charge mode leads to the checker-board charge-order pattern in real space. At the phase transition the mode diverges. Its dynamical properties are described and calculated in Ref. [18]. It is described based on the charge fluctuations which drive the system into the charge order-transition. For the calculations large-N and Lanczos diagonalization are used. The critical values for the charge order

3. Interplay of charge-order and superconductivity

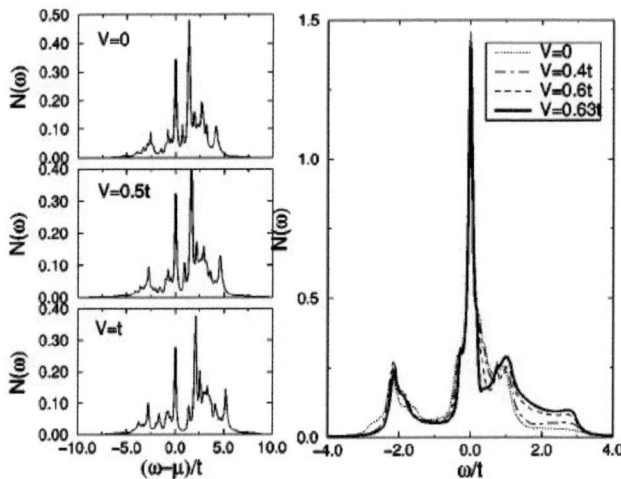

Figure 3.6.: A band from $\omega = t$ to $3t$ in the DOS develops as the charge-order transition is approached: Calculations are performed using Lanczos (left) and large-N (right) methods. A lower Hubbard band at $\omega 3t$ in the Lanczos results (at $\omega = -2t$ for large-N), the quasi particle peak at the Fermi surface, and the incoherent band around $\omega = t - 5t$ is seen where as the charge order transition is approached a pseudo gap below $\omega = t$ opens and spectral weight in the region around $\omega = t - 3t$ is enhanced. From [18].

transition using these methods are $V_c \approx 0.65t$ and $V_c \approx 2t$, respectively. (The discrepancy in the critical values for the different calculations models is a renormalization of the metallic part in the calculation that effectively changes the t [18]). The mode softens and gains weight as the charge order transition is approached. Its softening is the same described by both methods. That is shown in the inset where the peak position is given as a function of proximity to the charge order transition. Therefore the correlation values are normalized to the respective critical values V_c obtained from the different calculation methods. To derive the systems spectral densities and finally its total density of states (DOS), the self-energy is calculated. The interaction of the fermions with the collective mode (Fig. 3.5) is found to enhance the real and imaginary part of the self energy [18]. This leads to contribution to the DOS shown in Fig. 3.6. Compared to the DOS for the Hubbard model (like in half-filled systems) the band at $\omega = -2t$ is associated with the lower Hubbard band. It shows basically no change with different intersite Coulomb repulsion. Approaching the charge order transition the DOS at the Fermi surface is suppressed (better seen in the Lanczos results on the left) and a band in the $\omega = t - 3t$ region

3.2. Dynamics and spectral traces of metal, charge order, and superconductivity

is enhanced showing the strong influence of the nearest neighbor Coulomb interaction. To compare experimental results from optical measurements with the theoretical interaction model, the optical conductivity for increasing inter-site correlation was calculated using Lanczos diagonalization on a $L = 16$ cluster for $U = 20t$ as shown in Fig. 3.7. The critical value of correlations in that case is $V_c \approx 2t$. At $V = 0$, as discussed above, a Drude response is expected. Turning on correlations and approaching the charge order transition a broad band at about $\omega = 6t = 3V$ develops. This is due to the transition of charge carriers between different sites, where in case of occupied neighboring sites an energy of $3V$ has to be payed. That is the equivalent band to the Hubbard band at U in the case of half-filled systems, where the Coulomb repulsion to hop on an occupied site has to be overcome. Below that band a charge gap develops with increasing correlations. For correlations above the critical value this gap is fully developed. That signalizes the insulating charge order. On the metallic side close to the charge order transition, the system shows a pseudo gap behavior. The gap is filled in with a Drude response and a band around $\omega = 2t$. This band we call *charge fluctuation band* since it emerges from a resonant scattering of the quasi-particle scattering at the collective charge fluctuations mode. The presence of this band directly shows an interplay of the metallic and charge ordered state via the fluctuations. A simple coexistence would show a Drude response and a charge transfer band, only. The spectral weight of the charge fluctuation band basically represents the strength of the interaction of charge carriers with the charge fluctuations in the system. Summarizing, the optical spectrum in the charge ordered metallic state should show metallic properties in the Drude response and fingerprints of the charge fluctuations in the mid-IR band due to the temporary charge order patterns (Fig. 3.4 (d), and Fig. 3.7). At the same time the interaction between the charge carriers and the charge fluctuations is expressed in a charge fluctuation band that appears at low but finite frequencies.

3.2.4. Superconductivity

The peculiar charge-order metallic state is a pre-requirement for the possibility of superconductivity mediated via charge fluctuations [17, 18]. The interaction of the charge carriers with the charge fluctuations in this state is expressed by the H^{f-b} term in the Hamiltonian (3.2). It can be described as effective quasi particle potential.

Dynamical properties of the superconducting gap

As mentioned above and shown in more detail in the Appendix 14.1, the effective potential between quasiparticles is basically given by charge fluctuations and the non-double occupancy constraint. These are the same interactions which modulate the self-energy and are therefore responsible to the spectral weight shift of the charge transfer band and interaction features in the optical conductivity. These facts are closely related in the theoretical approach [18]. In that picture, more spectral weight at low but finite frequencies, especially the charge fluctuation band, means more effective interactions between the quasiparticles. In this framework, the same physics is involved in superconductivity.

3. Interplay of charge-order and superconductivity

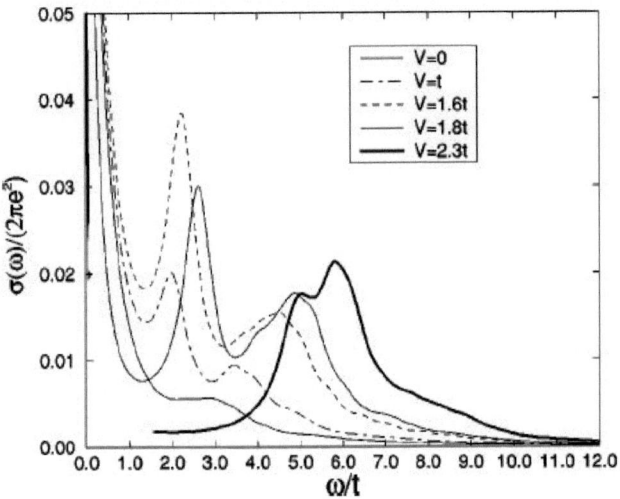

Figure 3.7.: Optical conductivity for different correlations V as the system goes through the charge order transition. Calculated from Lanczos diagonalization on a $L = 16$ cluster for $U = 20t$ using the extended Hubbard model on a square lattice. The critical value for the metal-insulator transition is $V_c \approx 2t$. Spectral weight is enhanced at low frequencies from $\omega = t - 6t$ as the transition is approached. Around $\omega = 2t$ the charge fluctuation feature develops. Above the transition the optical gap below the MIR band is present. From [18].

3.2. Dynamics and spectral traces of metal, charge order, and superconductivity

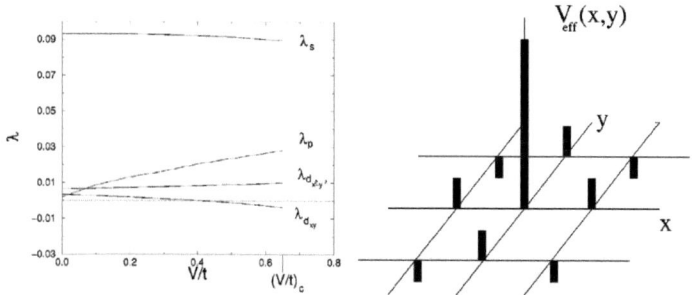

Figure 3.8.: (a) Effective coupling λ between the quasi-particles with increasing correlations V/t toward the critical value $(V/t)_c$ for the different pairing channels. (b) Effective potential $FT[V_{\text{eff}}]$ between the quasi particles in real space for the different symmetry channels. From [18].

That means, a strong charge fluctuation band together with the presence of a Drude peak should lead to high superconducting temperatures. Fig. 3.8 (a) shows the effective electron-electron coupling λ for different pairing channels on the square lattice. The d_{xy} channel becomes attractive for correlations $V/t \geq 0.4$ while the other channels stay repulsive. That the xy channel is more favorable is also seen in the real space sketch of the effective potential Fig. 3.8 (b). It shows the Fourier transform of the effective potential $V_{\text{eff}}(x,y) = FT\left[V_{\text{eff}}(q_x, q_y)\right]$ close to the critical value of correlation $(V/t)_c$. If one quasi particle is in the origin and a second quasi particle is placed there, it feels the large repulsion due to the on-site Coulomb interaction represented by the large positive bar. Along the x and y axes the potential also acts repulsive due to the intersite Coulomb repulsion. However along the diagonal, in xy-direction, the potential acts attractive represented by the negative bars. The attractive interaction along the diagonals is also in agreement with the underlying checker-board charge-order pattern which fluctuations are basically acting as the effective interaction.

Reflectivity and description of the superconducting gap

The presence of a superconducting state can be also seen in optics. The simplest description is within the BCS theory. Due to the formation of Cooper pairs the superconducting gap of 2Δ opens at the Fermi level. For frequencies within the gap the energy is not high enough to excite the charge carriers above the gap. There is no absorption of the Cooper pairs and therefore all light in that frequency range is totally reflected. The reflectivity jumps to 1 for $\omega < 2\Delta$ for T well below T_c as seen in Fig. 3.9. In the optical conductivity a gap opens below the superconducting gap frequency. The corresponding spectral weight is transfered to the delta peak at zero frequency. For gap symmetries different

3. Interplay of charge-order and superconductivity

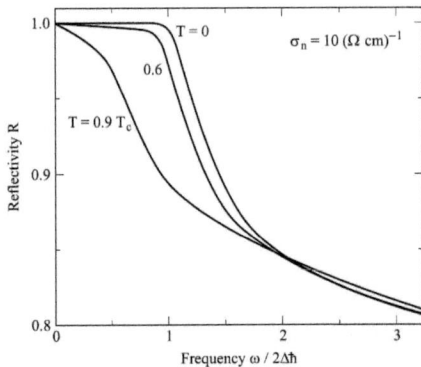

Figure 3.9.: Bulk refectivity of a superconducting metal as a function of frequency at different temperatures as indicated from the Mattis-Bardeen expression. From [22].

from BCS s-wave the behavior is more complicated due to k-dependent properties, the DOS states within the gap and the influence of the coherence factors [22].
The behavior of the gap can be described within BCS theory using the Mattis Bardeen formalism which gives the ratio of the superconductivity to the normal state conductivity. It is valid in the clean as well as in the dirty limit of superconductivity [90]. For the intermediate limit the extension of the formalism by Zimmermann and Brandt can be used [91]. It allows one to modulate or fit the behavior of conductivity and reflectivity based on the gap frequency, the normal state plasma frequency, and dc conductivity (or scattering rate) for different temperatures. The formalism also includes the clean and dirty limit.

3.3. Aspects of vibrational spectroscopy

Besides the pure electronic system, the vibrational system is influenced by the effective charges localized in the system. In addition to the lattice phonons, the molecular crystals show vibrations within the molecules. The modes due to the vibrations of the molecules on the lattice sites against each other are called lattice or phonon modes while the intramolecular vibrations of the molecules itself usually are called vibrational or molecular modes. Modes with a coupling between both are also possible [92, 93]. Compared to atomic systems the lattice phonons in (organic) molecular systems usually are expected at low frequencies in the FIR region, e.g. in BEDT-TTF salts below ≈ 200 cm^{-1}. This is due to the heavy masses of the molecules on the lattice sites and weak coupling strength between each other. The molecular vibrations are found in the same frequency range as the vibrations in the free molecules: The MIR region up to ≈ 3000 cm^{-1}. Depending

3.3. Aspects of vibrational spectroscopy

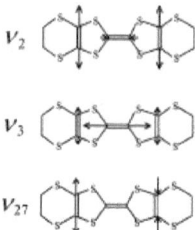

Figure 3.10.: Schematic view of the charge sensitive C=C double bond vibrational modes in BEDT-TTF molecules. The fully symmetric and Raman active ν_2 and v_3 modes; and the antisymmetric and IR active ν_{27} mode. From [100].

on the structure and composition of the anion there are also IR active vibrations within the anion molecules which are found then in the same spectral range.
Especially the molecular vibrations of the organic molecular crystals do strongly interact with the electronic system and therefore are altered in their properties. This is due to e-mv coupling to the molecular orbital [66, 92–99] as explained below (Sec. 3.3.2). Some modes can be used to probe the charge density directly on the molecule. These dependencies are explained in the following.

3.3.1. Charge dependent molecular vibrations

The C=C double bond vibrations highly dependent on the ionicity of the molecule and can be probed via IR- or Raman spectroscopy. Their center frequency can be used as a direct measure of the charge located on the lattice sites [74, 100–102]. Different ionicities of the BEDT-TTF molecule change the bond lengths within the molecule. This is measured (mainly for the central C-S and C=C bonds) in crystallographic x-ray to probe the charge on the molecule for BEDT-TTF [103–105]. The altered bond lengths change the force constants of the molecular vibrations. That leads to a frequency shift of the modes depending on the charge.
Usually the C=C double bond modes are measured. In the BEDT-TTF based systems these are the fully symmetric Raman active $A_g(\nu_2)$, $A_g(\nu_3)$, and the antisymmetric IR active $B_{1u}(\nu_{27})$ mode (Fig. 3.10). The frequency-charge dependence was investigated from the experimental [70, 100, 105–107] as well as from the theoretical [93, 97, 100, 105, 106] point of view. Depending on the charge per molecule the modes frequency shifts in a linear way, e.g. for the ν_2 and ν_{27} (Fig. 3.11). However the highly e-mv coupled ν_3 mode often shows a non linear frequency-charge dependence [74]. In the metallic state the charge density is equally distributed within the crystal in average. Therefore only one frequency should be observed per mode. If charge disproportionation or long range charge order sets in, the charge density is not distributed equally. More charge density is located on some and less at other sites. The modes split into different frequencies. From

3. Interplay of charge-order and superconductivity

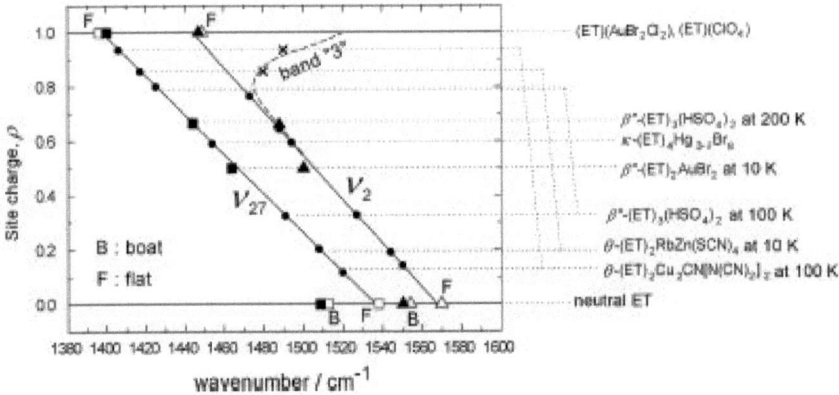

Figure 3.11.: Frequency dependence of the C=C double bond modes ν_2 (Raman) and ν_{27}(IR) due to different charge density on the molecule. Measurement values taken for different compounds. From [100].

the frequencies one can estimate the average charges on the different sites. The splitting translates then to the charge disproportionation between the sites. The modes are of Lorentzian shape (or Gaussian in case of disorder). A sharp mode corresponds to a quite localized state. A single frequency dominates. A fluctuating charge density leads to a broadening of the mode. Frequencies spread around the center frequency contribute due to less defined charge density on the molecule in time average. Fluctuations appear in the order of $10^{-12}s$ [108]. The investigations of the modes can be done either in Raman for the symmetric modes coupling to their polarization or in IR for the asymmetric ones coupling to their dipole moment. Often the symmetric ν_2 and ν_3 modes can be observed in the IR due to a symmetry break via e-mv coupling. In that case their center frequencies are already sifted due to the coupling and the frequency dependence of the charges is not linear any more. For the organic BEDT-TTF salts the IR-active ν_{27} mode has to be measured along the insulating direction because the mode is symmetry-forbidden in the in-plane direction.

3.3.2. Electron molecular vibration coupling

Many molecular vibrations in the conducting layer as well as in the anion layer can be treated as independent molecular vibrations like in a single molecule. They can be described as a driven oscillator and modeled by an Lorentzian oscillator. The same applies to the low-frequency lattice phonons. However, the structure of the molecular crystal enables the charge carriers to couple to the intramolecular vibrations. This possibility

3.3. Aspects of vibrational spectroscopy

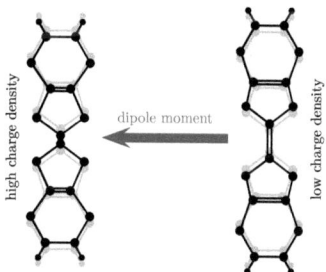

Figure 3.12.: Scheme of the e-mv coupling here for the Raman active ν_3 mode. The temporary dipole moment turns the mode also IR active.

depends on the actual molecular orbital and the symmetry of the vibrational mode. The coupling itself acts via an modulation of the energy in the HOMO orbital with the coupling constant $g_\alpha = \frac{\partial \epsilon}{\partial Q_\alpha}$, where ϵ is the HOMO energy and Q_α the vibrational normal coordinate [109–111]. As a result, an effective dipole moment appears as sketched in Fig. 3.12. In the presence of a symmetry break (in that case meaning the molecular sites are not on a symmetry center) the electron density is excited to oscillate with the phonon frequency along the sites: One site shows a high charge density while the next site a reduced charge density. Therefore a temporary dipole moment between them is present. The external field couples to it. This process of coupling a intramolecular vibration to the electronic background is called Electron-Molecular Vibration (e-mv) coupling. In 1D it can be strictly shown that only totally symmetric modes can couple [112]. In 2D a qualitative picture for totally symmetric modes can be drawn [95, 96].

E-mv coupled modes can rise significantly in spectral weight if they are coupled to a strong electronic band. It is not possible to use a Lorentz oscillators to fit them anymore. Here the Fano model [113] is a suitable way to describe a single energy level (the molecular vibration) which is coupled to a broad energetic background (the conduction band). A more detailed and exact theory for e-mv coupled phonons is given by a cluster approach from Yartsev [110] and Rice [109] taking into account the electronic part and the em-v coupled vibrations.

4. Investigated organic systems

Here we give an introduction into the known properties of the systems in focus of this thesis. The β''-(BEDT-TTF)$_2$SF$_5R$SO$_3$ family is in the center of the investigations and is a prime example to investigate the interplay of superconductivity and charge order. A similar system is the superconductor β-(EDT-TTF)$_4$[Hg$_3$I$_8$]$_{(1-x)}$. In terms of charge redistribution the α-(BEDT-TTF)$_2$NH$_4$Hg(SCN)$_4$, α-(BEDT-TTF)$_2$TlHg(SCN)$_4$, θ-(BEDT-TTF)$_2$RbZn(SCN)$_4$, and α-(BEDT-TTF)$_2$I$_3$ are in the focus of the vibrational analysis in this thesis. For the latter two compounds we address open questions in the phase transition.

4.1. The β''-(BEDT-TTF)$_2$SF$_5R$SO$_3$ family

The family of the (BEDT-TTF)$_2$SF$_5R$SO$_3$ organic compounds are based on experiments with large and discrete counter anions in order to synthesize organic superconductors [114]. They contain fully organic compounds, no metal atoms are present in the anions.

Main compounds of that series are achieved for SF$_5R$SO$_3$ anions with R=CH$_2$, CF$_2$, CHF, CHFCF$_2$, and CH$_2$CF$_2$ whereas the latter results in an organic superconductor with a transition temperature at around 5 K [83, 84, 114–116]. All systems are quarter-filled and are proposed to serve as a model system to investigate the interplay between charge order and superconductivity. Fig. 4.1 shows the position of the the two main compounds in the phase diagram for quarter-filled systems. The β''-(BEDT-TTF)$_2$SF$_5$CHFSO$_3$ is on the metallic side for all temperatures. The superconductor β''-(BEDT-TTF)$_2$SF$_5$CH$_2$CF$_2$SO$_3$ is in close proximity to the charge order transition. Therefore it is strongly influenced by the charge fluctuations. They mediate the attractive interaction that turns the system superconducting at lowest temperatures. The charge fluctuations are correlation driven and are also present at higher temperatures. There the system is in the charge ordered metal state. The other salts of the (BEDT-TTF)$_2$SF$_5R$SO$_3$ series are located beyond the charge order transition in the insulating phase.

So far attempts fail to synthesize a even higher correlated superconductor, e.g. using the slightly larger anion SF$_5$CHFCF$_2$SO$_3$. These system shows structural changes: It turns insulating at around 190 K due to anion disorder (Fig. 4.14(c)) [84, 117]. In the investigated compounds here electronic effects are dominant and effects due to structural disorder or phase transitions can be neglected. The superconductor shows a slight dimerization of the BEDT-TTF molecules at room temperature which holds on cooling. In the insulating compound it vanishes at decreasing temperature [116]. The dimerization does not influence the optical properties significantly. No opening of a dimerization gap is

4. Investigated organic systems

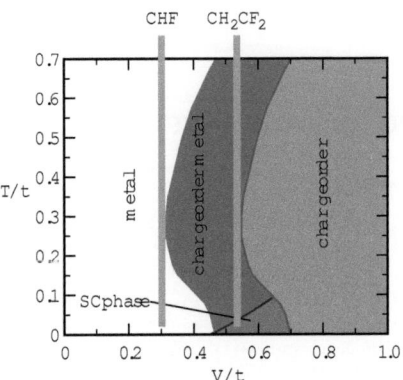

Figure 4.1.: Proposed position of the β''-(BEDT-TTF)$_2$SF$_5$CH$_2$CF$_2$SO$_3$ and the β''-(BEDT-TTF)$_2$SF$_5$CHFSO$_3$ in the phase diagram for quarter-filled two dimensional systems on a square lattice. All other compounds of the (BEDT-TTF)$_2$SF$_5$$RSO_3$ series become charge ordered due to structural effects in the anion layers.

observed [66, 84]. The anions with R=CH$_2$, CF$_2$, and CHF differ in their structural and physical properties. The CH$_2$ and CHF system are dimerized in their anion layers and form a β'' structure in the conduction layer. However, the CF$_2$ is monomeric in the anion layer and has a strongly dimerized stack of BEDT-TTF molecules. It is labeled as β'-system, which is basically the same structure as the β'' but with significantly increased dimerization of the BEDT-TTF stacks. The CH$_2$ anion system shows two differently charged donor molecules and favors electron localization. Both, the CF$_2$ and CH$_2$ become semiconducting/insulating [83]. The CHF system instead is claimed to show a pure metallic behavior down to lowest temperatures [84]. This large variety within the (BEDT-TTF)$_2$SF$_5$$RSO_3$ family already shows the strong influence of structure on their electronic properties: Especially the fact that the anion layer can induce disorder. However, the β''-(BEDT-TTF)$_2$SF$_5$CHFSO$_3$ and β''-(BEDT-TTF)$_2$SF$_5$CH$_2$CF$_2$SO$_3$ are both fairly ordered and similar in their anion layers. Therefore the differences between them are primarily due to the different degree of effective correlations. That makes them the ideal pair to investigate the interplay of charge order and superconductivity.

4.1.1. β''-(BEDT-TTF)$_2$SF$_5$CH$_2$CF$_2$SO$_3$

The β''-(BEDT-TTF)$_2$SF$_5$CH$_2$CF$_2$SO$_3$ was originally synthesized by electro crystallization in 1996 [114, 115] as the first fully organic superconductor with a $T_c \approx 5$ K.

4.1. The β''-(BEDT-TTF)$_2$SF$_5$RSO$_3$ family

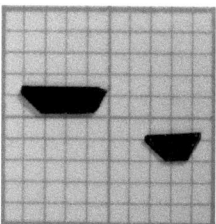

Figure 4.2.: The two largest and high quality β''-(BEDT-TTF)$_2$SF$_5$CH$_2$CF$_2$SO$_3$ single crystals used in this study on mm unit paper. Samples provided by J.A. Schlueter from Materials Science Division, Agronne National Laboratory, U.S.A.

To perform low frequency measurements of the optical response the availability of high quality single crystals is important. On the largest samples usable crystal faces were sized 2.5×1.2 mm^2 as shown in Fig. 4.2. The crystal surfaces were perfectly flat and free of kinks or steps.

Structure and bandstructure

The crystal structure (Fig. 2.7) contains BEDT-TTF cation layers separated by insulating SF$_5$CH$_2$CF$_2$SO$_3$ anion layers as typical for organic conductors (Sec. 2.2). The unit cell consists of four molecules arranged in the β''-motive (Fig. 4.3 (left)), where two of them are crystallographically independent (signed A and B), and the others centrosymmetric related to them (A', and B'). Crystallographic parameters in Tab. 4.1 for the isostructural compounds β''-(BEDT-TTF)$_2$SF$_5$CH$_2$CF$_2$SO$_3$ and β''-(BEDT-TTF)$_2$-SF$_5$CHFSO$_3$ are taken from Ref. [115, 116] and [83]. The molecules are oriented nearly parallel to each other forming loose stacks along the a direction ($t \approx 25 - 110$ meV). The orbital overlap to the adjacent stack in b direction ($t \approx 118 - 260$ meV) is larger than along the stacks. On cooling the overlap enlarges due to thermal contraction. The transfer integrals are given in Tab. 4.2 [116]. The ethylene end-groups of the BEDT-TTF molecule form hydrogen bonds with the outer fluorines and oxygen of the SF$_5$CH$_2$CF$_2$SO$_3$ anion. There is no intermoleculer overlap from the inner fluorine of the anion to the hydrogen of the cation. The anion has no chiral carbon atom and therefore the anion pocket is ordered (Fig. 4.14 (a)) [84]. The dimerization within the conducting plane is comparably low and is claimed to slightly increase on cooling based on optical investigations [66] while this is not seen in the structural properties [116]. X-ray structure did not show changes down to a temperature of 123 K (Tab. 4.1). Even a small reduction of dimerization is seen. Latest x-ray results do not show any changes even down to 20 K [118].

The electronic band structure can be described in a simple picture. Calculations restricted to the HOMO approach reveal a metallic band (Fig. 4.4 (a)). The room temperature optical response in the MIR frequency range is typical for BEDT-TTF salts

4. Investigated organic systems

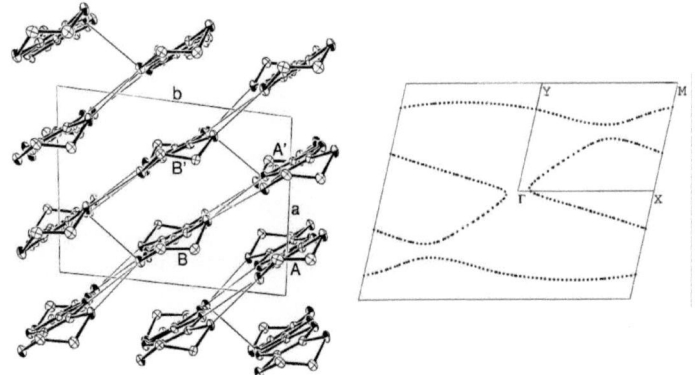

Figure 4.3.: The β''-type molecular arrangement of β''-(BEDT-TTF)$_2$SF$_5$CH$_2$CF$_2$SO$_3$ within the conducting plane (left) and the calculated Fermi surface using extended Hückel tight binding calculation (right). From [115]

T (K)	β''-(ET)$_2$SF$_5$CH$_2$CF$_2$SO$_3$ 298	123	β''-(ET)$_2$SF$_5$CHFSO$_3$ 298
density (g/mol)	1040.45		1008.43
a (Å)	9.260(2)	9.1536(6)	8.7965(2)
b (Å)	11.635(2)	11.4935(8)	11.7364(2)
c (Å)	17.572(5)	17.4905(12)	17.5763(4)
α (°)	94.69(3)	94.316(1)	95.893(1)
β (°)	91.70(1)	91.129(1)	90.178(1)
γ (°)	103.10(2)	102.764(1)	102.286(1)
V (Å^3)	1835.5(9)	1779.9(2)	1761.55(1)
Z	2		2
space group	$P\bar{1}$		$P\bar{1}$

Table 4.1.: Structural parameters for β''-(BEDT-TTF)$_2$SF$_5$CH$_2$CF$_2$SO$_3$ and β''-(BEDT-TTF)$_2$SF$_5$CHFSO$_3$. From [83, 115, 116]

4.1. The β''-(BEDT-TTF)$_2$SF$_5$RSO$_3$ family

	β''-ET$_2$SF$_5$CH$_2$CF$_2$SO$_3$		β''-ET$_2$SF$_5$CHFSO$_3$
T (K)	298	123	298
intrastack			
a	106	116	35
a'	25	58	95
b	69	124	86
b'	63	55	100
interstack			
c	260	259	283
c'	175	271	255
d	118	138	124
d'	130	154	138

Table 4.2.: Transfer integrals for β''-(BEDT-TTF)$_2$SF$_5$CH$_2$CF$_2$SO$_3$ and β''-(BEDT-TTF)$_2$SF$_5$CHFSO$_3$. Values are given in meV. (a, a'), and (b, b') are the alternating integrals within the $(A$-, $A')$- and $(B$-, $B')$-stack in Fig. 4.3; (c, c') the direct transfer integrals to the nearest neighbor, and (d, d') the integral along the diagonals to the next nearest neighbor in the adjacent stack. From [83, 116]

[84]. The whole electronic band spreads up to about 7000 cm^{-1}. A strong MIR contribution evidences the strong correlations in the system motivating to use a theoretical picture as described in Sec. 3. It suggests and justifies the description of the system within the (extended) Hubbard model. The hopping term t, setting the frequency axes for the correlation bands in the MIR, is in the same order of magnitude as the estimated Coulomb interactions for BEDT-TTF molecules. Therefore the treatment as correlated electron systems is appropriate and the choice of the theoretical parameters used in the theoretical approach is supported.

DC resistivity and Fermi surface

For β''-(BEDT-TTF)$_2$SF$_5$CH$_2$CF$_2$SO$_3$ the reports on DC resistivity at temperatures above the superconducting transition vary for different sources. Refs. [66, 115] report a semiconducting behavior down to a maximum at around 100 K and then a metallic behavior down to the superconducting transition but without showing data. On the other hand in Ref. [119] a metallic behavior down to 140 K followed by an increase in resistivity till 35 K is found before it enters the superconducting transition (Fig. 4.5 (a)). Most reports [84, 120, 121] claim a purely monotonic decrease in resistivity down to the superconducting transition but without showing data in the full temperature range. The superconducting transition measured in DC is shown in Fig. 4.5 (b). The only published data presenting purely metallic behavior is from the pressure dependent study [121] shown in Fig. 4.11 (a) (c.f. section about pressure dependence below). Temperature dependent ESR measurements are in line with the transport measurements as reported in Ref. [115]. The diversity of reports demonstrates the difficulties in measuring precise

4. Investigated organic systems

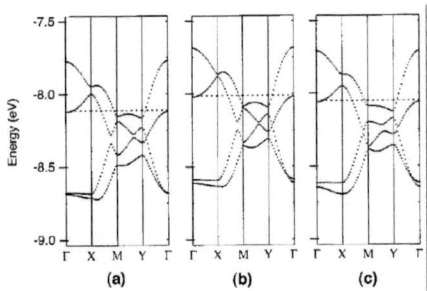

Figure 4.4.: Calculated band structure at room temperature using extended Hückel tight binding calculation. (a) β''-(BEDT-TTF)$_2$SF$_5$CH$_2$CF$_2$SO$_3$, (b) BEDT-TTF)$_2$SF$_5$CHFCF$_2$SO$_3$, and (c) β''-(BEDT-TTF)$_2$SF$_5$CHFSO$_3$. From [84].

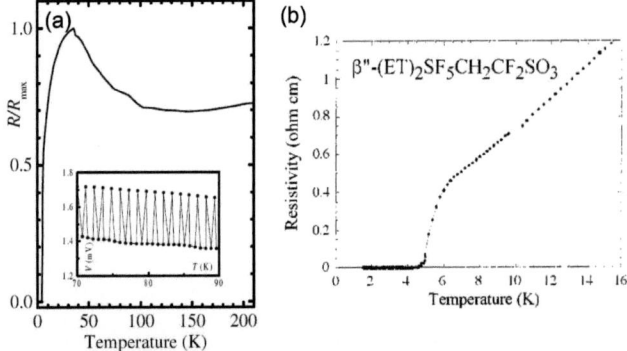

Figure 4.5.: (a) DC measurement of the interlayer resistance of β''-(BEDT-TTF)$_2$-SF$_5$CH$_2$CF$_2$SO$_3$. Metallic behavior down to 140 K is found. Then a maximum around 35 K appears before the system turns superconducting at 5 K. From [119]. (b) Low temperature DC measurement of β''-(BEDT-TTF)$_2$-SF$_5$CH$_2$CF$_2$SO$_3$. The superconducting transition appears around 5 K. From [84].

DC resistivity in organic crystals, especially for the in-plane direction. One carefully has to avoid cracks in the sample during the cooling procedure. They lead to contributions from the insulating directions of the system. Also in the earlier studies the crystal quality was not that high (as seen in a lower T_c). The strong anion influence in the other systems of the family shows that slight defects in the anions layer of β''-(BEDT-TTF)$_2$SF$_5$CH$_2$CF$_2$SO$_3$ can lead to disorder and explain the semiconducting behavior at intermediate temperatures. However, for all resistivity measurements a metallic behavior is found right above the superconducting transition at $T_c = 5 \pm 0.4$ K.

The calculated Fermi surface (Fig. 4.3 (right)) contains 2D hole pockets and 1D sheets [115]. Experimentally the Fermi surface was mapped out using de Haas-van Alphen (dHvA) [122, 123], Shubnikov-de Haas (SdH) [120, 123], angular-dependent magneto resistance oscillations (AMROs) [124, 125], and millimeter-wave magneto-conductivity [126] but no evidence for the 1D sheets could be found. Further, the size of the measured Fermi surface in the experiments is only about a fifth (5 % of the 1st Brioullin zone) of the expected one (25.4 % of the 1st Brioullin zone). The presence of the two-dimensional Fermi surface proves that the system is metallic. The fact that only a small part of the Fermi surface is found suggests that only a small fraction of the carriers contributes to the coherent transport. Most of the carriers are localized or ordered due to correlation effects. No Fermi surface nesting is seen. The influence of the correlations is evident in the cyclotron and band effective mass extracted from Shubnikov-de Haas measurements as $m_c = 1.9 m_e$ and $m_b = 3.9 m_e$ respectively. In review [35] Wosnitza points out that β''-(BEDT-TTF)$_2$SF$_5$CH$_2$CF$_2$SO$_3$ is a perfect quasi two-dimensional system: No coherent interlayer transport is evident, and dHvA oscillations perfectly fit a two-dimensional metal [127]. Summarizing, the transport properties of the superconductor show a metallic state carried by a small number of carriers. That is in perfect agreement with a charge ordered metallic state due to charge fluctuations close to the charge order transition as described in Sec. 3.

Optical properties

In first optical studies, Dong et al. [66] performed IR reflectance measurements on mosaics samples and found the presence of a huge mid-IR band which they assigned to interband transitions. The room temperature optical conductivity is shown in Fig. 4.6. The broad mid-IR band fits to the simple band picture drawn in the HOMO-LUMO approach. From partial sum rule on optical conductivity along the high conducting b direction an effective mass of $m_b^* = 1.6 m_e$ was extracted [66, 84]. Down to lowest temperatures no distinct Drude peak was found, as seen in Fig. 4.7. It shows the temperature dependence of the optical response. That is in discrepancy to the metallic behavior in DC resistivity and the Fermi surface characterization. They prove the existence of a two-dimensional Fermi surface. The missing Drude response in the framework of Ref. [66] is likely because it is below their accessed frequency range of about 50 cm^{-1} (whereas within this thesis the low frequency response down to 10 cm^{-1} could be measured). Instead a red shift and grow of spectral weight of the mid-IR band is observed which they attributed to charge localization and interdimer interaction of slightly increasing

4. Investigated organic systems

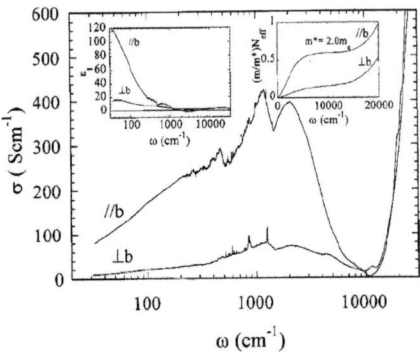

Figure 4.6.: Optical conductivity of β''-(BEDT-TTF)$_2$SF$_5$CH$_2$CF$_2$SO$_3$ obtained from IR reflectance measurements at room temperature. From [66].

dimerization. The band rising in the 200 cm^{-1} region is not addressed and the spectral weight there is assigned to phonons, only. Here we investigate that band as charge fluctuation band which drives the interplay in the charge ordered metallic state. The phonon spectra are identified and analyzed in detail. The assignment of the IR modes in the range from 240-2983 cm^{-1} within the conducting plane can be found as table in Ref. [66]. The enhanced interaction due to e-mv coupling (c.f. Sec. 3.3.2), increased dimerization, or small structural changes is suggested. Raman studies of the vibrational modes at room temperature reveal no charge disproportionation; just a single non-split ν_3 mode is detected [116]. This is in agreement with the metallic character of the system at room temperature. Low temperature Raman measurements were not performed so far.

Following the idea of the interplay of charge order and superconductivity the former results can be reinterpreted and extended. The missing Drude response and the strong localized behavior seen in optics together with the small Fermi surface suggest that the system is close to a charge order transition. In addition its transpoert properties are metallic at all temperatures. That makes it an ideal candidate for a charge ordered metallic state [69]. At low temperatures such a state is proposed to support superconductivity mediated by charge fluctuation [17, 18]. The presence of a low frequency band in the optical spectrum also allows the interpretation within a charge fluctuation picture as motivated in Sec. 2 and discussed in Sec. 3.

4.1. The β''-$(BEDT$-$TTF)_2SF_5RSO_3$ family

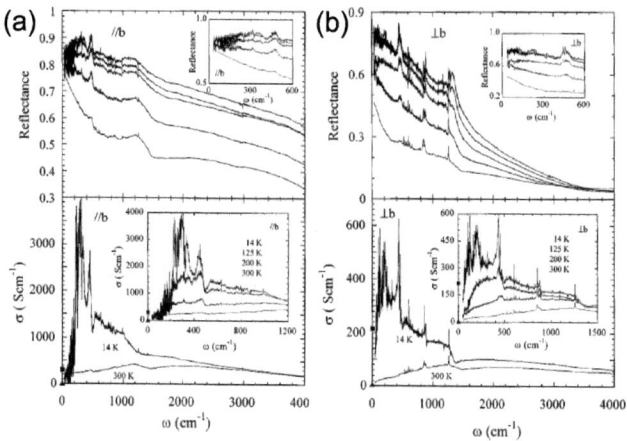

Figure 4.7.: Temperature dependent optical reflectivity and conductivity of β''-(BEDT-TTF)$_2$SF$_5$CH$_2$CF$_2$SO$_3$ obtained from IR reflectance measurements (a) perpendicular and (b) along the stacking direction. From [66].

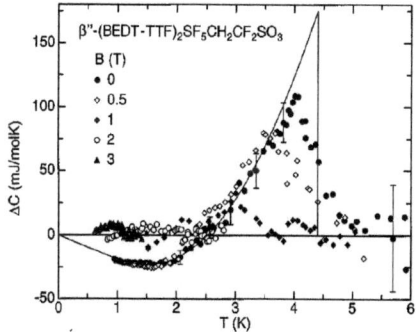

Figure 4.8.: Specific heat difference ΔC between the superconducting and normal C in β''-(BEDT-TTF)$_2$SF$_5$CH$_2$CF$_2$SO$_3$. The solid line represents BCS fit in the strong coupling limit to the $B=0$ data. From [128].

4. Investigated organic systems

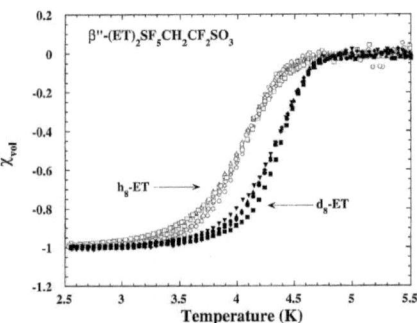

Figure 4.9.: Deuteration dependence in β''-(BEDT-TTF)$_2$SF$_5$CH$_2$CF$_2$SO$_3$ investigated by ac-susceptibility. An inverse isotope effect is found in contradiction to a simple BCS like behavior. From [129]

Superconductivity: Characterization with specific heat and deuterated samples

To characterize the superconducting state measurements of specific heat, on deuterated systems, and pressure dependence have been performed. The specific heat jump at the superconducting transition Fig. 4.8 can be well described using strong coupling BCS theory [128]. The coupling constant is extracted to be $\lambda = 1.1$ which is in agreement with the extracted effective mass enhancement from the SdH experiments. An upper critical field for the superconducting state is around $B_c = 3.5$ T, the in-plane coherence length is about 10 nm.

Based on deuterium substitution of the ethylene end groups in BEDT-TTF a 0.25 K increase of the superconducting transition temperature T_c is found in ac-susceptibility measurements (Fig. 4.9) [129]. That is in contradiction to the expectations from BCS theory. Changes due to the isotope effect are predicted to be about -0.04 K. The origin of this inverse effect is suggested to be related to a shortened C-D bond length compared to the C-H bond. This leads to a softer lattice and increases the electron phonon coupling. From that findings a strong phonon influence to the superconductivity is expected. Alternatively the transition could be influenced by intramolecular vibrations, or it results from highly uniaxial negative pressure derivatives as found in thermal expansion experiments (Fig. 4.10) [130]. These lattice effects are not included in the electronic picture presented in Sec. 3. But phonons and molecular vibrations are good candidates to expand the model of the electronic interplay between the fluctuations and coherent charges.

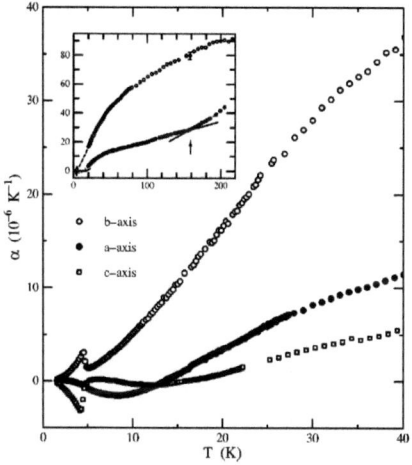

Figure 4.10.: Thermal expansion coefficient in $\beta''\text{-(BEDT-TTF)}_2\text{SF}_5\text{CH}_2\text{CF}_2\text{SO}_3$ along the crystal axes. The in-plane expansion coefficients along a and b direction are shown up to high temperatures as an inset. From [130].

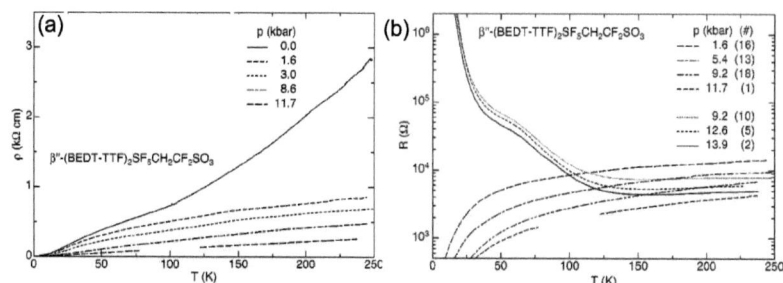

Figure 4.11.: Temperature dependent DC measurement of $\beta''\text{-(BEDT-TTF)}_2\text{-}$ $\text{SF}_5\text{CH}_2\text{CF}_2\text{SO}_3$ for different pressure. Up to 12 kbar (a) the system becomes more metallic with decreasing T_c before the behavior turns insulating (b) for higher pressure. From [121].

4. Investigated organic systems

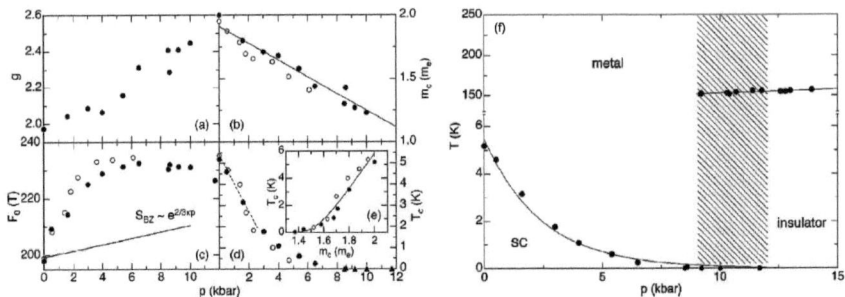

Figure 4.12.: β''-(BEDT-TTF)$_2$SF$_5$CH$_2$CF$_2$SO$_3$ pressure dependence of (a) the electron g factor, (b) the effective cyclotron mass m_c, (c) the SdH oscillation frequency ($\Theta = 0$), and (d) the superconducting transition temperature T_c in the metallic phase. The pressure dependent studies lead to a $P - T$ phase diagram as proposed in (f). In the shaded area both metallic and insulating behavior can be stabilized experimentally. From [121].

Pressure dependence of the superconducting state

In pressure dependent experiments the influence of the correlations can be verified. Like chemical pressure the physical pressure changes the transfer integrals and therefore the effective correlations. In β''-(BEDT-TTF)$_2$SF$_5$CH$_2$CF$_2$SO$_3$ a pressure induced phase transition to an insulating state is found [121]. This is counterintuitive since the pressure suggests to make the system more metallic.

Indeed, a small pressure makes the interlayer transport more metallic due to an enhanced overlap between the layers as measured with DC transport (Fig. 4.11). On the other hand no indications for an three-dimensional Fermi surface are found: The interlayer transport stays incoherent and the system stays quasi-two-dimensional [121, 127]. The superconducting coupling parameter λ obtained from cyclotron effective mass, and the transition temperature T_c are lowered (Fig. 4.12 (b) and (d)). That indicates reduced effective correlations. The system is pushed towards the metallic phase.

But at high pressure, above 12 kbar, the system surprisingly turns insulating for temperatures below 153 K. On the metallic side the Fermi surface topology stays unchanged as followed by AMRO. But SdH measurements (Fig. 4.12 (c)) find that the Fermi surface first increases faster than can be explained just by the lattice compression. Then it decreases close to the insulating transition which reveals a reconstruction of the band structure [121]. That suggests a strong pressure influence on many-body parameters. It is expressed in an enhanced band mass due to electron-electron or electron-phonon coupling. Summarizing, a phase diagram as in Fig. 4.12 (f) is proposed from the findings. The metal-insulator transition shows a hysteresis of about 3 kbar. That points to structural changes. This is in agreement with SdH experiments which show a band

4.1. The β″-(BEDT-TTF)₂SF₅RSO₃ family

Figure 4.13.: Typical β″-(BEDT-TTF)₂SF₅CHFSO₃ single crystal under the microscope. Crystal faces are highly tilted and small and therefore need to be exactly aligned under the IR microscope. Samples provided by J.A. Schlueter from Materials Science Division, Agronne National Laboratory, U.S.A.

structure reconstruction preceding the phase transition. On the other hand neutron scattering experiments do not show significant changes in the crystallographic structure [131]. However, these pressure induced changes are not due to electronic correlations, only. Ref. [132] shows that pressure changes the slope of $\rho = \rho_0 + AT^2$ proportional to the cyclotron effective mass to the power of 6 (m_c^6). A scaling to the square (m_c^2) is expected for a pure electronic origin. That opposes the Fermi liquid nature of the T^2 behavior. Further, the comparison with the CHFCF₂ compound shows that slight structural changes can lead to insulating phases. This holds also as a possible explanation in the low temperature upturn of resistivity in some metallic CHF samples below 6 K. Summarizing, the pressure dependence is in line with the interplay picture of the correlation driven charge fluctuations. The system becomes more metallic with increasing pressures. The metal-insulator transition at high pressure opposes this trend but it is likely due to structural changes in the anions or strong renormalization of the electron-phonon coupling. The latter proposes a strong influence of the vibrational system to the electronic properties.

4.1.2. β″-(BEDT-TTF)₂SF₅CHFSO₃

The β″-(BEDT-TTF)₂SF₅CHFSO₃ is the metallic compound of the (BEDT-TTF)₂-SF₅RSO₃ family and is less investigates so far. Single crystals (Fig. 4.13) here are of poorer quality compared to the superconducting sister compound. They grow with maximum dimensions of $1 \times 1.8 \times 0.3$ mm³. Usable crystal surfaces are sized 0.5 mm in diameter on non flat sample edges. That makes it necessary to use microscopy or gold evaporation techniques to measure the optical properties (see Sec. 6).

Structure and bandstructure

The β″-(BEDT-TTF)₂SF₅CHFSO₃ metal is isostructural to the β″-(BEDT-TTF)₂-SF₅CH₂CF₂SO₃ superconductor with conducting layers of β″-motive separated by the

4. Investigated organic systems

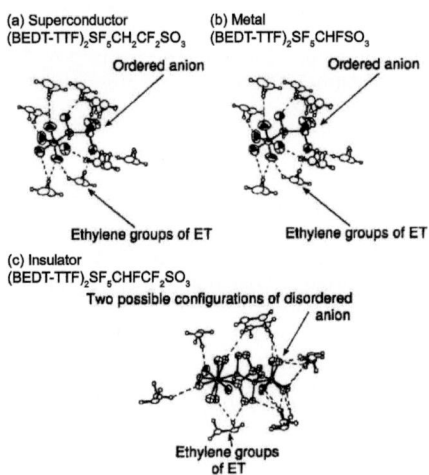

Figure 4.14.: Structure of the anion pocket in the (BEDT-TTF)$_2$SF$_5$$RSO_3$ family. (a) In the case of the superconductor R=CH$_2$CF$_2$ the anion pocket is fully ordered. (b) The metal R=CHF it is still fairly ordered. (c) The insulator R=CHFCF$_2$ shows disorder in the anion pocket due to two possible anion configurations. From [117].

4.1. The β''-(BEDT-TTF)$_2$SF$_5$RSO$_3$ family

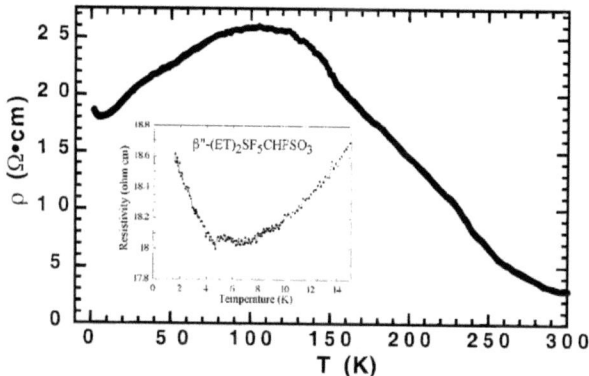

Figure 4.15.: Low temperature dc measurement of β''-(BEDT-TTF)$_2$SF$_5$CHFSO$_3$ reveals an activated behavior down to 100 K, the metallic behavior down to 6 K. From [83]. The inset shows the low temperature region with the resistivity upturn below 6 K. From [84].

insulating anion layer. The same symmetry considerations as for the superconductor apply. The structural parameters are given in Tab. 4.1. The transfer integrals ($t \approx 35 - 100$ meV intrastack, and $t \approx 124 - 285$ meV interstack) (Tab. 4.2) are slightly higher than in the superconducting sister compound. While the structure of the conducting layer is similar to the one in the β''-(BEDT-TTF)$_2$SF$_5$CH$_2$CF$_2$SO$_3$ superconductor, the anion structure in the the metal β''-(BEDT-TTF)$_2$SF$_5$CHFSO$_3$ slightly differs. The SF$_5$CHFSO$_3$ anion possesses a chiral carbon atom but compared to the metal/insulating CHFCF$_2$ system there is only one fluorine having some intermolecular bonds to the BEDT-TTF donor molecules (Fig. 4.14). Therefore the anion pocket is still ordered. This is believed to support the metallic behavior down to low temperatures [84] while disordered pockets seem to lead to a metal-insulator transition.

DC resistivity and Fermi surface

In the DC characteristics for β''-(BEDT-TTF)$_2$SF$_5$CHFSO$_3$ there is a discrepancy in the literature. It is expected to be metallic behavior down to lowest temperatures but at the same time below 6 K an upturn in resistivity of unknown origin is found [83, 84]. A slight disorder in the anion pocket could be a reason for that. Nevertheless, no full range data of the purely metallic behavior is available; the published data (Fig. 4.15) shows semiconducting behavior: An increase of resistivity from room temperature down to 100 K before the metallic behavior sets in [83]. The calculated Fermi surface and band structure (Fig. 4.4 (c)) are similar to the one of the superconductor, as expected

4. Investigated organic systems

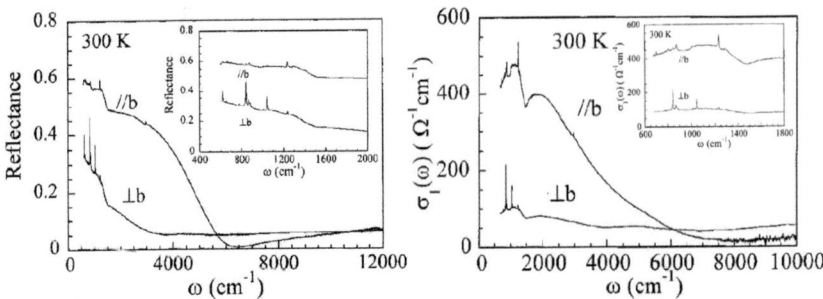

Figure 4.16.: Room temperature IR reflectance measurement and extracted optical conductivities of β''-(BEDT-TTF)$_2$SF$_5$CHFSO$_3$ perpendicular and along the stacks. The insets show a close up of the vibrational structure in the 600-1800 cm^{-1} range. From [84].

due to the isostructural arrangement of the BEDT-TTF molecules in the conducting plane. They also support the metallic picture.

Optical properties

The optical properties along the principal axes of the β''-(BEDT-TTF)$_2$SF$_5$CHFSO$_3$ are investigated at room temperature and down to 600 cm^{-1}, only. The response (Fig. 4.16) shows a broad mid-IR band, in agreement with the simple band structure (Fig. 4.4 (c)). From partial sum rule on optical conductivity along the high conducting b direction an effective mass of $m_b^* = 1.7 m_e$ was extracted [84]. However, no distinct Drude peak was found [84]. Temperature dependent optical measurements were not performed so far. On vibrational properties, anion modes at 840, 1048, and 1240 cm^{-1} as well as the ν_{60} molecular vibration of BEDT-TTF at 860 cm^{-1} were identified.

Other experimental results

The charge per molecule was estimated to be +0.47e and +0.53e based on x-ray bond-length measurements [83]. Room temperature Raman measurements show a ν_3 vibrational mode at 1471 cm^{-1} [83]. The mode does not split (Fig. 4.17 (c)). That indicates only a small charge redistribution between the molecular sites. In contrast to that, the insulating sister compounds Figs. 4.17 (a) and (b) show a split mode due to the charge order.

In room temperature ESR characterization, a linewidth ΔH between 26-38 G and g value from 2.002-2.012 depending on orientation are extracted [83]. The maxima in both, ΔH and g, appear when the magnetic field is along the direction of the central C=C bond of the BEDT-TTF molecule, the minima with the field perpendicular to the molecular

4.1. The β''-(BEDT-TTF)$_2$SF$_5$RSO$_3$ family

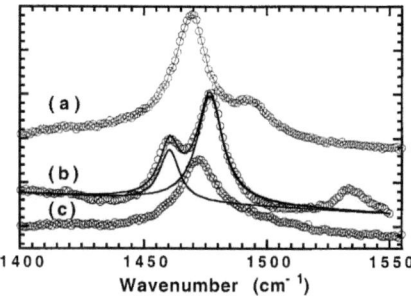

Figure 4.17.: Room temperature Raman measurement of the ν_3 vibrational mode in (a) β'-(BEDT-TTF)$_2$SF$_5$CF$_2$SO$_3$, (b) β''-(BEDT-TTF)$_2$SF$_5$CH$_2$SO$_3$, and (c) β''-(BEDT-TTF)$_2$SF$_5$CHFSO$_3$. From [83].

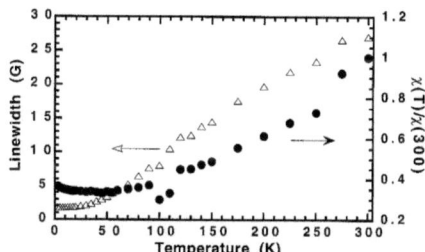

Figure 4.18.: Temperature dependent ESR linewidth and relative spin susceptibility of β''-(BEDT-TTF)$_2$SF$_5$CHFSO$_3$ for the magnetic field oriented perpendicular to the molecular (ab)-plane. From [83].

75

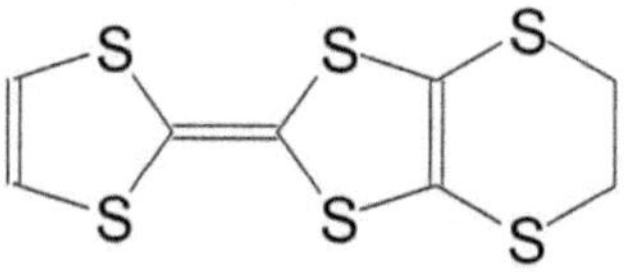

Figure 4.19.: The ethylenedithiotetrathiofulvalene (EDT-TTF) molecule. From [133].

plane. The temperature dependent ESR linewidth and relative spin susceptibilities are given in Fig. 4.18. The linewidth decreases on cooling to lowest temperature. Also the spin susceptibility decreases from room temperature down to 100 K. That supports the semiconducting behavior in DC-transport [83]. Around 100 K the slope flattens which indicates a metallic behavior. Below 40 K the spin susceptibility slightly increases again. That would indicate electron localization. But it is also pointed out that the properties depend a lot on experimental parameters and sample quality [83].

Summarizing, the β''-(BEDT-TTF)$_2$SF$_5$CHFSO$_3$ is the ideal sister compound to the β''-(BEDT-TTF)$_2$SF$_5$CH$_2$CF$_2$SO$_3$. Both compounds are isostructural and for high quality samples basically no anion disorder influences the electronic properties. Therefore all differences between the compounds can be attributed to their different effective correlations. The electron system in the β''-(BEDT-TTF)$_2$SF$_5$CHFSO$_3$, which stays metallic, is assumed to be less correlated compared to the superconductor β''-(BEDT-TTF)$_2$SF$_5$CH$_2$CF$_2$SO$_3$.

4.2. The β-(EDT-TTF)$_4$[Hg$_3$I$_8$]$_{(1-x)}$ organic superconductor

The β-[(EDT-TTF)$^{+0.5}$]$_4$[Hg$_3$I$_8$]$^{-2}_{(1-x)}$ superconductor is a quasi two-dimensional quarter-filled system similar to the BEDT-TTF based charge transfer salts. The stoichiometry is 4:1 with an ionicity of 2 for the Hg complex. The building block molecules in the conducting plane are the asymmetric EDT-TTF molecule (Fig. 4.19). Crystals are provided by Institute of Problems of Chemical Physics, Russian Academy of Sciences, Chernogolovska, Russia.

Structure and DC characterization

The molecular arrangement in the conducting plane has a β-motive [134, 135]. The crystal structure is shown in Fig. 4.20. On the same batch of crystals, slight deviations from the stoichiometry are found. There are two descriptions in literature to describe these deviations: In the first papers it is expressed as (EDT-TTF)$_4$Hg$_{3-\delta}$I$_8$ [133, 136] while a later paper investigates the stoichiometry more thoroughly and explains β-(EDT-TTF)$_4$[Hg$_3$I$_8$]$_{(1-x)}$ [134]. The paper shows detailed crystallographic data and structural arrangement of the anion and cation layers for slight stoichiometric deviations. Based

4.2. The β-(EDT-TTF)$_4$[Hg$_3$I$_8$]$_{(1-x)}$ organic superconductor

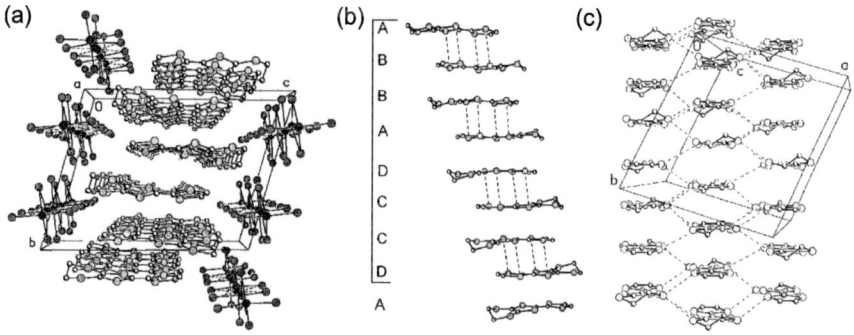

Figure 4.20.: (a) The crystal packing of β-(EDT-TTF)$_4$[Hg$_3$I$_8$]$_{(1-x)}$ along the a-axis, (b) stacking sequence, and (c) interstack interactions. From [134].

on that, some samples show a metal-superconductor transition around 8 K while others turn insulating [133, 134, 136]. The DC characteristics for the different cases are shown in Fig. 4.21 on the left. Electron-probe micro-analysis of the stoichiometry [134, 136] reveals a non-stoichiometry of $x = 0.03$ for the superconducting samples (a). Samples with $x = 0.02$ become superconducting under a slight pressure of about 0.3 kbar (b) and (c). The stoichiometric sample (d) undergoes a metal to insulator transition below 30-35 K. Applying pressure on this just lowers the onset of the metal-insulating transition. As possible explanation for this effect, changes in the order/disorder of the anion chains is given. That is similar to the influence observed in the (BEDT-TTF)$_2$SF$_5$$RSO_3$ family. The defect sections of short anion subchains of β-(EDT-TTF)$_4$[Hg$_3$I$_8$]$_{(1-x)}$ are shown in Fig. 4.21 on the right. These defects are assumed to give the distortion to the conducting layer which supports the superconductivity [134].

Magnetic field dependence of resistivity

Magnetic field dependence of the resistivity was measured parallel and perpendicular to the conducting plane (Fig. 4.22) [136]. The step in the dependencies with the field applied parallel to the plane (a) was assigned to the possibility of a coexistence of different superconducting states. In the normal conducting state the crystals revealed a small negative magnetoresitance [136]. The temperature dependent critical fields estimated from the recovery of the expected magnetoresitance are given in Fig. 4.23.

Optical properties

V. Semkin from the Ioffe Ph. Tech. Inst. in St. Petersburg has measured the EDT-TTF response of the β-(EDT-TTF)$_4$[Hg$_3$I$_8$]$_{(1-x)}$ in the mid-IR at room temperature and 5 K

4. Investigated organic systems

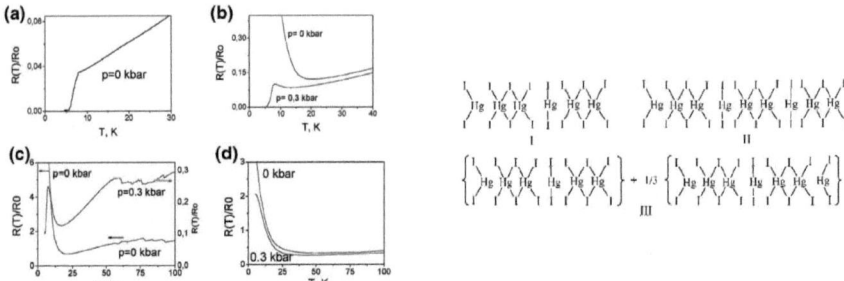

Figure 4.21.: Temperature dependent resistivity for single crystals (left) of β-(EDT-TTF)$_4$-[Hg$_3$I$_8$]$_{(1-x)}$ with different stoichiometry deviations. The corresponding sub-chain configuration is shown on the right. (a ↔ III, b ↔ II, c ↔ I). (d) is the stoichiometric sample. From [134].

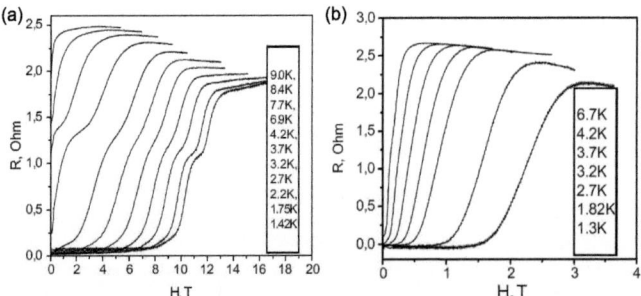

Figure 4.22.: Magnetoresistance in β-(EDT-TTF)$_4$[Hg$_3$I$_8$]$_{(1-x)}$ measured at different temperatures with the magnetic field (a) parallel and (b) perpendicular to the conducting plane. From [136].

4.3. The α-(BEDT-TTF)$_2$MHg(SCN)$_4$ family family

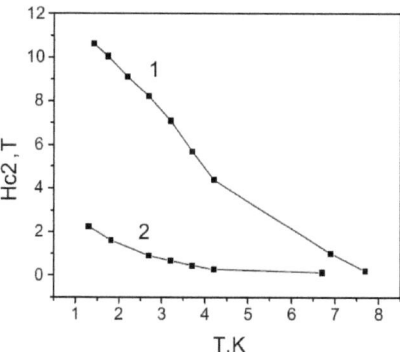

Figure 4.23.: Critical field in β-(EDT-TTF)$_4$[Hg$_3$I$_8$]$_{(1-x)}$ for the magnetic field (1) parallel and (2) perpendicular to the conducting plane as function of temperature. From [136].

(unpublished). Basically the optical response is similar to the one known from quarter filled BEDT-TTF charge transfer salts.

Summarizing, the quarter-filled β-(EDT-TTF)$_4$[Hg$_3$I$_8$]$_{(1-x)}$ is a good candidate to investigate the general behavior of the interplay between charge order and superconductivity. The existence of superconducting and insulating samples, in our case even within one experimental batch, suggest the system to be close to the charge order transition. That agrees nicely with its high $T_c \approx 8$ K. Also the fact that the insulating samples become superconducting under pressure shows the tendency of pushing the sample back to the more metallic side of the phase diagram. In addition, the strong influence to the anion layer reveals a strong similarity to the (BEDT-TTF)$_2$SF$_5$RSO$_3$ family.

4.3. The α-(BEDT-TTF)$_2$NH$_4$Hg(SCN)$_4$ organic superconductor and the α-(BEDT-TTF)$_2$TlHg(SCN)$_4$ organic metal

To investigate high correlated metals in the proximity of the charge order transition, α-(BEDT-TTF)$_2$NH$_4$Hg(SCN)$_4$ and α-(BEDT-TTF)$_2$TlHg(SCN)$_4$ are chosen. They are a well known superconductor and metal compound out of the α-(BEDT-TTF)$_2$MHg(SCN)$_4$ family containing salts with M=K, Rb, Tl, and NH$_4$ [137, 138]. For the different metal ions the the [MHg(SCN)$_4$]$^-$ has the same valence but different volume. That applies the chemical pressure and changes the effective correlations. However, the relative strength of the different overlap integrals along the different directions also changes [79]. The NH$_4$-system has a typical two-dimensional corrugated cylindrical Fermi surface [80],

4. Investigated organic systems

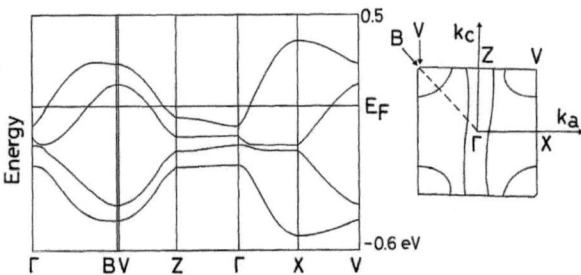

Figure 4.24.: Band structure and Fermi surface of α-(BEDT-TTF)$_2$NH$_4$Hg(SCN)$_4$. A quasi-two-dimensional metal is found. From [79].

in agreement with the systems metallic character. In contrast to β''-(BEDT-TTF)$_2$-SF$_5$CH$_2$CF$_2$SO$_3$, a pair of open electron-like Fermi surfaces exits along the c-direction. Further there is a closed hole-like Fermi surface around the V point. They are shown in Fig. 4.24 together with a tight binding based band structure calculation [139]. The density of states ($N(E_F)$) was calculated to increase with the anion volume (K<Rb<NH$_4$). Then T_c is expected to increase, too [79]. But the NH$_4$-system is the only compound that become superconducting, the other salts enter a density wave state at low temperatures [81, 140]. That might be due to a nesting of the one-dimensional parts of the Fermi surface. The absence of these in the (BEDT-TTF)$_2$SF$_5$RSO$_3$ family does not give this possibility there.

The optical response of α-(BEDT-TTF)$_2$MHg(SCN)$_4$ was already presented in Sec. 2.2.4. The systems are close to the charge order transition in the charge-ordered metallic stae. But the proposed location within a correlation dependent phase diagram (Fig. 2.23, based on optics) is opposing the tendencies seen in e.g. the β'' family. It also opposes the phase diagram from theoretical models. Reason for that might the rather complicated structure of α-salts. Here they will be compared in the framework of vibrational spectroscopy to clarify to what extend there are parallels to β'' systems and where are the differences. Main focus will be the degree of charge redistribution which is expected in the charge ordered metallic state.

4.4. The θ-(BEDT-TTF)$_2$RbZn(SCN)$_4$ and α-(BEDT-TTF)$_2$I$_3$ charge order compounds

θ-(BEDT-TTF)$_2$RbZn(SCN)$_4$ and α-(BEDT-TTF)$_2$I$_3$ are chosen as high correlated systems. They show a phase transition into an insulating charge ordered state. The structure of θ-(BEDT-TTF)$_2$RbZn(SCN)$_4$ is close to the β''-type. As motivated in the introduction it represents a square lattice which can be considered as the first approxi-

4.4. The θ-(BEDT-TTF)$_2$RbZn(SCN)$_4$ and α-(BEDT-TTF)$_2$I$_3$ charge order compounds

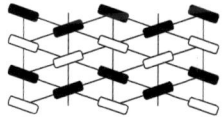

Figure 4.25.: Horizontal stripes charge order pattern in α-(BEDT-TTF)$_2$I$_3$ or θ-(BEDT-TTF)$_2$RbZn(SCN)$_4$ below the metal insulator transition at 135 K or 190 K respectively.

mation for the triangular lattice of the β''-systems. In comparison to the α-(BEDT-TTF)$_2$MHg(SCN)$_4$ family, the θ-(BEDT-TTF)$_2$MM'(SCN)$_4$ systems are located closer to the charge ordered phase [141]. For θ-(BEDT-TTF)$_2$RbZn(SCN)$_4$ experiments reveal that a charge disproportionation already develops in the metallic state before the transition [75, 142, 143]. That is similar to the behavior which is expected for the β''-systems. In the α-phase, α-(BEDT-TTF)$_2$I$_3$ is the only system that exhibits a fully charge-ordered insulating state [61, 144]. There a non-fluctuating charge order is present already above the insulating state [145].

4.4.1. α-(BEDT-TTF)$_2$I$_3$

The α-(BEDT-TTF)$_2$I$_3$ structure with its slightly dimerized stacks with the so called α-I$_3$ type packing motive as described in Sec. 2.2. The system is one of the most prominent and well studied example of a charge order based metal insulator transition. The conductivity drops orders of magnitude at $T_{MI} = 135$ K. The optical properties were presented in Sec. 2.2.4. The phase transition is caused by charge-order and is heavily investigated in ^{13}C-NMR [64, 145–148] and Raman [73, 74] studies. The charge order pattern is horizontal stripes oriented perpendicular to the stacks (Fig. 4.25). The charge disproportionation in the above studies is estimated to $\Delta\rho \geq 0.6e$. Some of the NMR studies [145, 148] even show a charge disproportionation which is present above the transition. It increases with decreasing temperature as sketched in Fig. 4.27. The unit cell does not change. However, the metal-insulator transition stays sharp. X-ray studies reveal no structural changes down to 120 K while measurements at 20 K give some hints for a structural transition at temperatures below metal insulator transition temperature [149].

4.4.2. θ-(BEDT-TTF)$_2$RbZn(SCN)$_4$

The θ-(BEDT-TTF)$_2$RbZn(SCN)$_4$ structure is similar to the one of α-(BEDT-TTF)$_2$-I$_3$ but without dimerization along the stacks. Therefore the unit cell is only half the size. A metal-insulator transition takes place at 190 K. A horizontal stripe charge-order develops. In contrast to α-(BEDT-TTF)$_2$I$_3$, the transition in θ-(BEDT-TTF)$_2$-RbZn(SCN)$_4$ is accompanied by a structural transition. A slight dimerization along the

4. Investigated organic systems

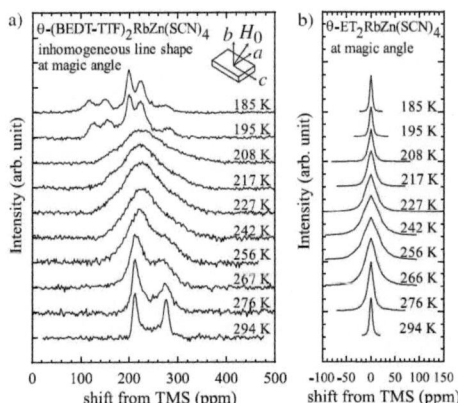

Figure 4.26.: NMR study of θ-(BEDT-TTF)$_2$RbZn(SCN)$_4$ showing the (a) inhomogeneous and (b) homogeneous lineshape for the field tuned within the ab-plane. The broadening of the lines above $T_{CO} = 195$ K indicates the fluctuations present before the charge order transition. From [142]

stacks appears and leads to a translational offset between the stacks and a rotation of the molecules [150] which is important for the horizontal stripe pattern. Also a slight dimerization perpendicular to the stacks sets in. The unit cell doubles and ends up like in the α-(BEDT-TTF)$_2$I$_3$. Above 190 K a short range fluctuating 3-fold charge order forms studied in NMR [142, 151, 152] before it enters the charge ordered insulating state [63, 77]. Fig. 4.26 shows the temperature dependence of the inhomogeneous and homogeneous line shape of ^{13}C-NMR for θ-(BEDT-TTF)$_2$RbZn(SCN)$_4$. The two peaks at room temperature correspond to two crystallographic independent sites in the BEDT-TTF molecule. The broadening of the lines above the transition temperature at $\approx 195\ K$ indicates the fluctuating charge-order. In the insulating state a combination of a broad and sharp peaks appears. As in the α-(BEDT-TTF)$_2$I$_3$, the charge-order pattern turns into long range stripe charge order. This is sketched in Fig. 4.28. The structural change at the metal insulator transition leads to a doubling of the unit cell. That makes the unit cell similar to the one of α-(BEDT-TTF)$_2$I$_3$.

4.4. The θ-(BEDT-TTF)$_2$RbZn(SCN)$_4$ and α-(BEDT-TTF)$_2$I$_3$ charge order compounds

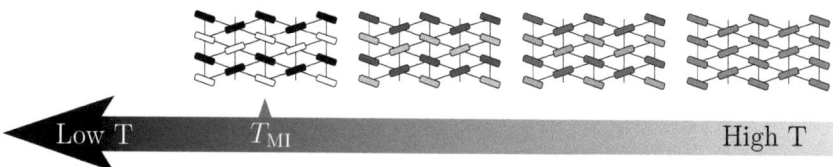

Figure 4.27.: Sketch of the temperature dependent behavior of α-(BEDT-TTF)$_2$I$_3$. For high temperatures (room temperature) the charges are equally distributed (0.5e per site). With decreasing temperature the charge disproportionation increases until the system undergoes the metal-insulator transition.

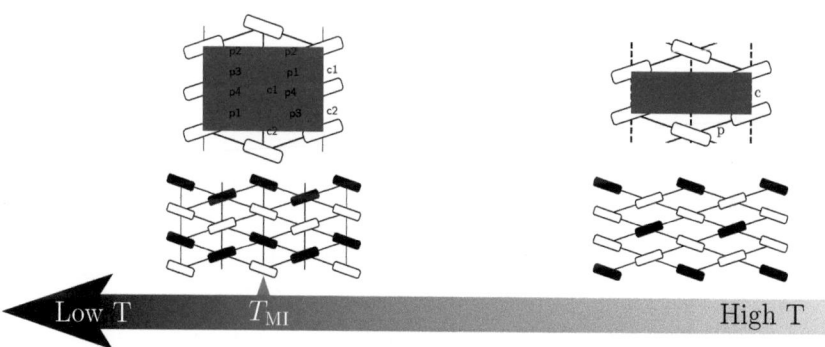

Figure 4.28.: Sketch of the temperature dependent behavior of θ-(BEDT-TTF)$_2$-RbZn(SCN)$_4$. Above the transition a 3-fold fluctuating charge order is present. Along with the metal-insulator transition a structural transition occurs leading to a doubling of the unit cell.

83

5. Aims of the thesis: Interplay of CO and superconductivity

Aim of the thesis is to understand the interplay of charge-order and superconductivity. Organic conductors and superconductors with different degree of correlations show metallic, charge ordered insulating, and superconducting ground states. They are used as model systems to investigate the corresponding degree of charge order and its influence to the optical properties. A successful interpretation of their spectra allows to draw a picture of the electronic interactions in the materials. This can be compared then to other known correlated materials and theories.

5.1. Characterizing the degree of charge order in a metallic, charge ordered and superconducting phase

As motivated in the materials section for the quarter-filled two-dimensional systems a strong relationship between superconductivity and charge-order is suggested. A charge-ordered metallic state is proposed in vicinity to the charge order transition. At low temperatures this state could support superconductivity. To characterize it, the transport properties of the different systems are investigated: DC-resistivity is measured using four- and two-contact methods. Further, the microwave properties are investigated using cavity perturbation technique. One main interest is the different degree of charge order for a metal, a superconductor, and a charge ordered insulating system. In molecular crystal that is directly detectable via its vibrational properties. The center frequency of the C=C double bonds modes are a direct probe of the average charge per molecular site.
Therefore an IR setup has to be realized that allows to measure the temperature dependence of the vibrational properties along the insulating direction of a single crystal. For this purpose an IR microscope has to be equipped with a micro cryostat. The alignment inside the cryostat turns out to be of major importance so that a micro positioning stage has to be developed. That allows to align the samples with respect to a reference mirror. This method opens the possibility to perform temperature dependent reflectance measurements in the microscope on tiny and non flat or irregular shaped samples.

5.2. Influence of the charge order to the optical properties

The optical response of the different systems measures the changes of the electronic properties. Of special interest are systems close to the charge order transition. There a charge ordered metallic phase due to charge order fluctuations is proposed. These systems tend to become superconducting on cooling to low temperatures. The metallic properties give rise to a Drude response while charge order leads to the formation of a charge transfer band and the opening of an optical gap. Optical investigations should also reveal whether there is just a simple coexistence of the metallic and charge ordered state within the crystal or if there is a real interaction between them. In the case of a coexistence the phases would compete and one feature would increase or decrease at the expense of the other one. In the interplay case there should be an additional feature due to the interaction between the two phases. The existence of such a feature has to be proven and its properties to be analyzed. An extended Drude description of the coherent response gives the possibility to trace interactions of the quasi particles with collective excitations. A theory based on the charge fluctuation model predicts an interaction feature (charge fluctuation band) as a consequence of the quasi particles interacting via the charge fluctuations in the system.

To obtain the electronic properties of the systems, the reflectance of the systems in a broad frequency (THz-VIS) and temperature range (2-300 K) is measured. A Kramers Kronig analysis allows to extract the optical conductivity as response function. Especially the reflectance measurements in the THz range have to be pushed to lower frequencies. There the wavelength is close to the scattering limit for the small single crystals. For tiny and poorly shaped crystals, e.g. the available crystals of the organic metal β''-(BEDT-TTF)$_2$SF$_5$CHFSO$_3$, the possibility to measure reliable temperature dependent properties using an IR microscope has to be developed and its low frequency limit has be pushed to the FIR as far as possible. Therefore the micro-sample-stage is used which is constructed for the measurement of the vibrational properties.

5.3. Overview picture about the interplay of charge fluctuations and superconductivity

The interpretation of the spectra allows to compare the electronic properties to other correlated systems. Further it can be checked to what extend existing theories can explain the observed physics. Is possible to draw a direct connection from the charge fluctuations in the metallic state to superconductivity? Does such an interplay purely act on the electronic level? Strong coupling effects to the lattice system are found in the investigated and also in related systems. Is it possible to to implement and understand this in the framework of superconductivity?

Part II.

Experiments

Chapter 11

Experiments

6. Experimental techniques

To investigate the organic conductors and superconductors a series of different techniques is used. Measuring the optical response in a broad frequency range probes the dynamical properties of the electron system. FTIR spectrometers cover the VIS/NIR range down to the FIR. THz spectroscopy accesses the frequency range below that. In this regime two different techniques are used: BWOs as a tunable coherent wave THz source, and a Martin-Puplett interferometer. In addition, vibrational spectroscopy directly probes the charge order in the systems: Raman and IR microscopy. Standard DC measurements and microwave investigations based on the cavity perturbation technique probe the transport properties.

6.1. Optical measurements

Optical reflectance spectroscopy is a direct probe of the dynamical properties of the electronic system: The light couples directly to the dipole moment of the electronic states. At low frequencies, in this case in the THz and FIR regime below $100 cm^{-1}$, the dynamics of the coherent-carrier response is measured while at higher frequencies the excitation of localized carriers or interband transitions takes place.

6.1.1. Broadband FTIR spectroscopy in the VIS/NIR-FIR range

Two different FTIR spectrometers cover the broad range of frequencies in these studies. Their working principle is based on the Michelson interferometer (Fig. 6.1). The incoming beam splits in two at the beam splitter and travels along different arms. Then it is overlapped at the beam splitter, again. The light path in one arm is varied by a moving the end mirror. The temporal coherence of the incoming light allows to measure the constructive and destructive interference of the two beams depending on the time delay between them. The Fourier transform of the resulting interferogram gives the frequency spectrum of the incoming light. A detailed description of the technical implementations and mathematical algorithms used in FTIR spectroscopy can be found in [153, 154].

The reflected energy spectrum is measured for the sample and then is referenced to the spectrum of a 100% reflecting mirror. Al or Ag mirrors are used in the NIR and VIS regime and Al or Au mirrors in the FIR and THz region. The mirrors reflection coefficients are corrected based on literature [155]. A uncertainty of less than 0.2% absolute reflectivity is reached by averaging multiple scans. A large variety of different light sources, beam splitters, and detectors span a broad frequency range as given in Tab. 6.1

6. Experimental techniques

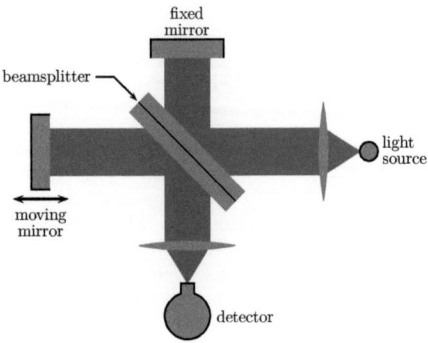

Figure 6.1.: Sketch of a Michelson interferometer as used in the Bruker 66. Beam splitter and source can be replaced depending on the required frequency range (c.f Tab. 6.1).

Range (cm^{-1})	Source	Beam Splitter	Materials	Detector
10-60	Hg lamp	50 μm Mylar	Polypropylene, Polyethylene	Si bolometer 1.4 K
30-700	Hg lamp	6 μm Mylar	Polypropylene, Polyethylene	Si bolometer 4.2 K, DTGS
500-8000	Globar	Ge on KBr	KBr, ZnSe, KRS-5	MCT 77 K, DTGS
3000-10000	W bulb	CaF$_2$	CaF$_2$, Quartz	InSb 77 K

Table 6.1.: Combination of source, beamsplitter, detector, and materials useful for optical elements as windows and polarizers for the various frequency ranges of FTIR spectrometers.

6.1. Optical measurements

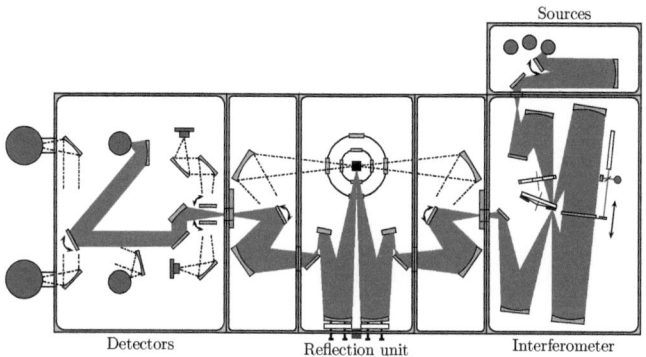

Figure 6.2.: Overview of the Bruker 113 spectrometer with extended detector chamber. The interferometer is of Genzel-type which enables high resolution measurements. The extended detector chamber and automatic source and beamsplitter changers allow to measure with two bolometers in the FIR using the same alignment and without breaking the vacuum during the same temperature run.

In addition, to measure polarization and temperature dependence suitable polarizer materials and window materials for optical cryostats are available.

High frequency measurements are done on a Bruker 66. It is based on a Michelson interferometer (Fig. 6.1) where the laser interferometer to determine the mirror position is inside the main interferometer for the optical path. The IR signal itself gives white-light position. This enables a high accuracy of the mirror position and therefore high precision up to frequencies in the VIS/UV. The MIR regime (500-8000 cm^{-1}) is measured in the Bruker 66 and in a Bruker 113. The low frequencies are performed on a Bruker 113. It is based on a Genzel-interferometer where the moving mirror changes both light arms at the same time. See Fig. 6.2 for the interferometer design and the optical path in the experiment and detector chambers. This interferometer has the double change in the optical path with the mirror movement compared to the Michelson design. That allows very high resolution measurements with a very compact interferometer design. The laser interferogram to measure the mirror position and the whitelight interferogram are separate outside the main interferometer for the IR signal. That leads to some limitations at very high frequencies. Therefore we use the Bruker 66 in that frequency range. As advantage the Bruker 113 has an extended detector chamber. One can switch between two bolometers, a 4.2 K broad range (30-700 cm^{-1}) and a 1.4 K pumped bolometer (20-150 cm^{-1}), using the same optical alignment in one temperature run.

For the FTIR measurements (including the vibrational spectroscopy) a resolution of 1 cm^{-1} in the whole frequency range up to 8000 cm^{-1} is used. The higher frequencies

6. Experimental techniques

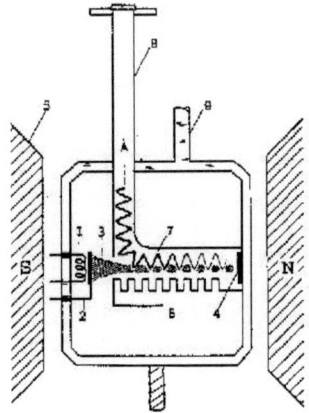

Figure 6.3.: Working principle of a backward wave oscillator (BWO). 1-heater; 2-cathode; 3-electron beam; 4-anode; 5-permanent magnet; 6-decelerating system; 7-electromagnetic wave; 8-cooling water flow.

are measured with a resolution of 8 cm^{-1}. The reflectivity curves were taken in runs sets of 256 single scans each on sample and reference mirror. Up to 10 curves are averaged for each temperature. 5 of them are reproduced with a completely new alignment of the whole setup. A reproducibility was reached better than 0.2% absolute reflectivity.

6.1.2. THz spectroscopy

To access the frequency range below the FTIR setup, two different sets of THz setups are used. They allow us to measure the low-lying features like the Drude response and cover the frequency range of the expected superconducting gap of the β''-(BEDT-TTF)$_2$-SF$_5$CH$_2$CF$_2$SO$_3$ superconductor $T_c \approx 5.4$ K.

Quasi optical coherent wave spectrometer

A quasi optical setup measures absolute reflectance values. It is equipped with backward wave oscillators (BWO) as coherent-wave THz source. These are tunable sources covering a frequency range from 1-40 cm^{-1}. The working principle [156] is sketched in Fig. 6.3. High voltage accelerates electrons which focused and guided within a static magnetic field. They become grouped in bunches by passing over a comb-like deceleration system. This creates an electromagnetic wave which travels in the opposite direction (backward-wave oscillator) and exits into free space through an oversized waveguide. The output frequency depends on the electron velocity and is controlled by the acceleration voltage.

6.1. Optical measurements

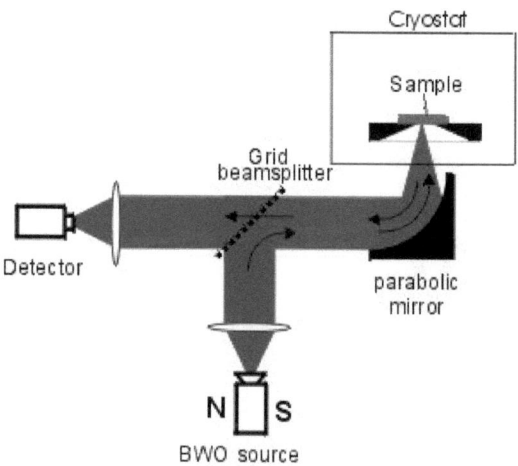

Figure 6.4.: THz reflectance setup. Polarizing beam splitter grids and teflon lenses are used. The focusing into the cryostat is done with an 90° off axis mirror.

The measurement setup is given in Fig. 6.4. It contains the BWOs as frequency tunable source. The beam is reflected from beam splitter (a suitable dense wire grid to split the beam 50/50, the passing light is lost) and is focused inside a gas exchange cryostat (1.75K-300K). The back reflected light passes the beam splitter (the reflected light is lost) and is focussed to the detector, which can be either a golay cell or a pumped 1.4 K Si bolometer [157]. For polarization dependent measurements an additional polarizer (very dense wire grid) sits in front of the focusing mirror.
In the THz regime the frequency resolution chosen is about 0.05 cm^{-1}. Sample and reference are measured in two subsequent scans. Up to 20 measurements are averaged per temperature step.

Light-pipe based Martin Puplett type interferometer

The temperature dependent opening of the superconducting gap and its magnetic field dependence in the THz regime are investigated with a different approach. Relative changes of the sample reflectivity with temperature and magnetic field are measured. The setup is the TeslaFIR^2-spectrometer at the THz group of the KBFI at Tallin (Estonia). It is based on the Sciencetech SPS-200 spectrometer with a Martin-Puplett interferometer [158] as sketched in Fig. 6.5. A linear polarized beam (polarizer P1) is split at beam splitter D1 into two beams perpendicular polarizes to each other. The two beams overlap at a second beam splitter D2, again. The resulting beam is elliptically polarized depending

6. Experimental techniques

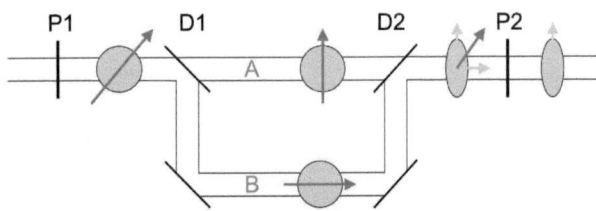

Figure 6.5.: Working principle of a Martin Puplett style polarizing interferometer. A beam is polarized at 45 degrees at the polarizer P1 and split into two perpendicular to each other polarized beams (A and B) at a polarizing beam splitter D1. The phase difference between the two beams overlapped at the polarizing beam splitter D2 results in an elliptically polarized beam being analyzed with the polarizer P2. From [158]

on the phase difference between the two beams. Its ellipticity changes periodically with the phase difference between the two beams. Using an analyzer (P2) polarized parallel or perpendicular to the incident polarization results in intensity changes as for the Michelson interferometer. The advantage of this setup is that the efficiency depends on the quality of the polarizers, only. High quality polarizers are available in the THz regime while in that frequency range FTIR spectrometers are limited by strong absorption features in the beam splitter.

The measurement setup of the TeslaFIR^2-spectrometer is sketched in Fig. 6.6. The light from the interferometer is guided through light pipes into the cryostat, a 12 T magnet system. Two ^{3}He bolometers inside the cryostat access the frequency range from 2-250 cm^{-1}. Measurements are taken on 2 samples and a reference mirror (to correct for relative system changes). The mirror is made of a 20% Au/ 80% Ag alloy. It shows only small changes in resistivity and magneto-resistance on cooling and in high magnetic fields. For polarization dependence an additional polarizer is placed directly in front of the sample.

6.2. Vibrational spectroscopy

In vibrational spectroscopy there are two main methods. Raman techniques access fully symmetric modes via their polarization. The coupling is proportional to the Raman tensor which describes the polarizability of the system. Its a two photon process via an excitation of a virtual state where the light is scattered at the phonon. On the other hand the infrared method probes the asymmetric modes which show a dipole moment the light can couple to. In this case it is a direct excitation with one photon.

6.2. Vibrational spectroscopy

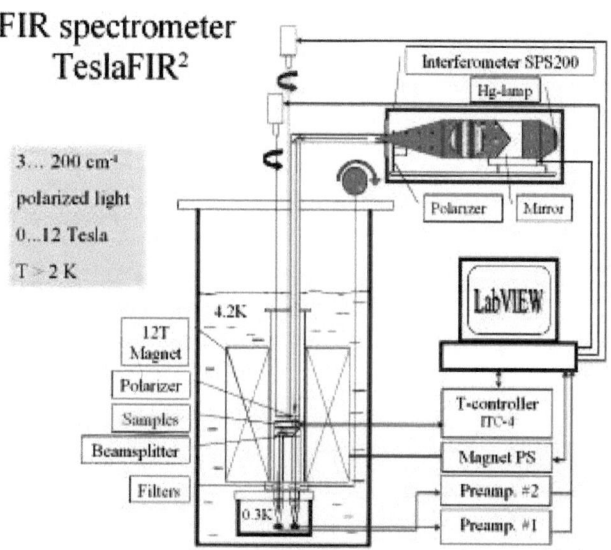

Figure 6.6.: The TeslaFIR²-setup at the KBFI, Tallinn (Estonia). It uses a mercury arc lamp as the light source in a polarizing (Martin-Puplett) interferometer SPS-200. Two He³ bolometers pumped to 0.3 K are used as detectors. The samples can be cooled down to 2.5K in a 12T magnet (Faraday and Voigt configurations possible).

6. Experimental techniques

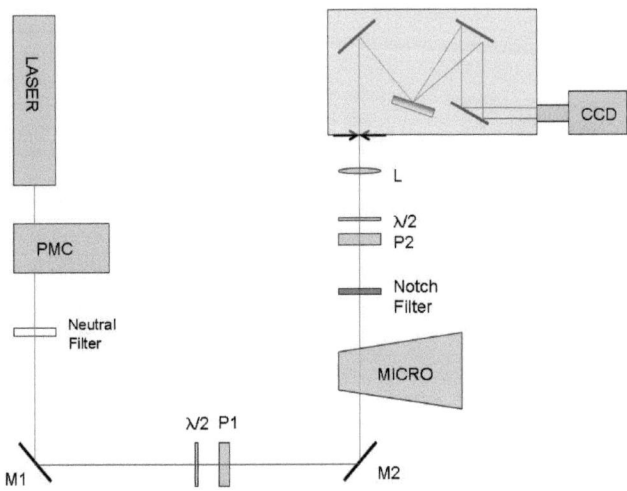

Figure 6.7.: Raman microscopy setup for high energy Raman scattering at the Dep. Chimica G.I.A.F. and INSTM-UdR at Parma university. Optical components: (1) Tunable laser, (2) PMC: prism monochromator removes the laser plasma lines, (3) $\lambda/2$-plate and polarizer P1 control the incident light polarization, (4) Microscope focuses the laser light on the sample and collects the Raman scattered light, (5) Notch filter rejects the reflected and elastically scattered light, (6) analyzer P2 controls the scattered light polarization, (7) $\lambda/2$-retarder rotates (when needed) the light polarization in the direction of highest sensitivity, (8) Grating specially splits the Raman scattered light frequency dependent to the channels of a CCD detector.

6.2.1. Raman microscopy

A standard Raman setup is sketched in Fig. 6.7. It consists of a tunable laser (Coherent Kr-ion laser) which acts as incident light source. Changing the excitation frequency optimizes the absorption of light in the crystal by tuning the excitation to an absorption band of the system. In the present case of organic systems its tuned to the interband-tansitions because the Raman scattered light scales with ν^4 of the incident frequency. A microscope focusses the light on the sample. The alignment of the sample is tilted to the incident beam so that most of it is scattered out and only the Raman scattered light is collected in the objective.

6.2. Vibrational spectroscopy

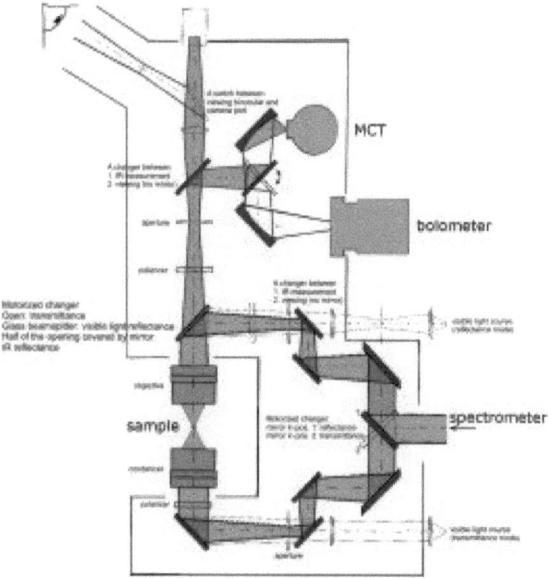

Figure 6.8.: Light path of the Bruker Hyperion IR micoscope attached to the Bruker66 FTIR spectrometer. Mirrors are used to guide and focus the beam.

6.2.2. IR microscopy

For IR vibrational spectroscopy a FTIR spectrometer (section 6.1.1) (Bruker 66) equipped with an IR microscope is used. The infrared beam is guided through the microscope using mirrors. A Schwarzschild objective (15×, opening NA 0.4, shaded center NA 0.17) focusses down to spot sizes from 20-250 μm. Here, vibrational spectroscopy is performed along the insulating direction of the system. Therefore the measurement is on the thin faces of the organic salts which allows spot sizes of about 40-80 μm. Usually this sample faces are not perfectly flat. To ensure the parallel alignment of sample surface and reference mirror a new micro sample stage was designed allowing independent change of the sample from the mirror alignment (Fig. 6.10).

6. Experimental techniques

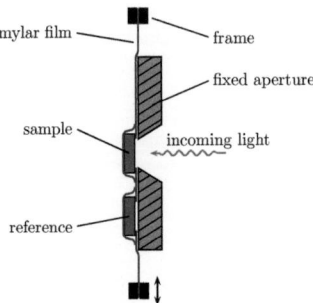

Figure 6.9.: Sectional view of the sliding aperture frame. Sample and mirror in the movable frame are placed in between thin mylar foils. The frame is moved up an down to place the sample or mirror behind the fixed aperture. To the aperture opening holes in size of the aperture are cut into the mylar foil through which the light can pass.

6.3. Optical low temperature systems and sample mounting

To achieve reliable absolute reflectance values, the optical pass should be the same for the sample and reference measurement. Depending on the setup, different sets of cryostats and sample holders are available.

6.3.1. Gas exchange cryostat with sliding aperture technique

The gas exchange cryostat enables to have optimal temperature control over the sample since it is placed inside a helium atmosphere. Also it is possible to operate the system at temperatures well below the liquid helium temperature by pumping on the cooling helium. Temperatures down to 1.75 K could be reached in the optical setups. Temperature stability was achieved to ±0.001 K.

Inside the gas exchange cryostat the measurements were performed using a sliding aperture technique sketched in Fig. 6.9. An aperture is spatially fixed and the sample and reference are exchanged behind that. This design keeps the same optical light path for the sample and reference measurement. To ensure the parallel alignment of sample and reference surface, the crystal and mirror are put on a piece of thin mylar which has holes in the size of the aperture to pass the light. Moving the mylar film back and forth enables to slide the sample or the reference in front of the aperture.

6.3.2. Cold finger cryostat for gold evaporation

To check the absolute values achieved with the sliding aperture method, gold evaporation measurements are done in addition. In this method, first all temperature steps are

6.3. Optical low temperature systems and sample mounting

Figure 6.10.: Construction drawing of the sample holder unit for the micro cryostat. The table in the middle can be tilted around two axes to align the sample face to be measured parallel to the reference mirror represented as gray box.

measured on the sample. Then a golden film is evaporated on the sample surface and all temperature steps are measured again. The evaporated gold film acts as perfect mirror. The setup used was built up and characterized within Ref. [159]. For high temperatures a good agreement between the methods is found while the gold evaporation could not be used at lower temperatures. For the β''-(BEDT-TTF)$_2$SF$_5$CH$_2$CF$_2$SO$_3$ the golden film peeled off due to the different shrinking of the crystal and the gold surface. A further drawback of the gold evaporation technique is that the sample is glued with carbon paint to a cold finger cryostat because a proper gold evaporation operates in the vacuum only. Therefore a good thermal coupling of the sample to the cold finger has to be ensured by gluing. Temperatures down to 5-10 K can be reached, only. Another disadvantage is that the sample has a golden film evaporated on it after measurement. Depending on sample robustness only one or just a limited number of measurement can be performed. Especially for organic charge transfer salts, which cannot be polished, typically one or two runs can be done, only. An advantage is that even for small samples the whole surface and also rough surfaces can be measured. The sample surface itself becomes a perfect mirror with exactly the same shape and steps after the gold evaporation.

6.3.3. Micro cryostat with micro positioning unit for IR microscopy

In the case of the microscopy measurements the sample and mirror are to be aligned parallel to each other and perpendicular to the incoming beam to ensure the same optical path in the sample and reference measurement. Both are measured in the focus spot. To be able to deal with uneven and non flat sample surfaces, a sample stage has to be constructed for that purpose. Required parameters are described in the following. The the stage (Fig. 6.10) is constructed in a way that it replaces the flat sample stage of the cold finger micro cryostat and has dimensions of 30 mm in diameter and 20 mm in hight. The flat mirror sits on a fixed post. The sample stage is made adjustable by tilting in two

6. Experimental techniques

directions to align the sample surface parallel to the mirror surface. The alignment has to be done under the microscope. That requires to make the adjustment skews accessible from top. Therefor the sample plate hangs in an outer ring which can be tilted around its center between two posts. The tilting is performed by pressing down the ring against a spring. The ring hold an inner plate which can be tilted on its center but on an axis perpendicular to the one of the outer ring. The tilting here is performed by the inner post that presses a ball to the inner plate to fix the tilting hight. Thermal contact to the cold finger system is reached by soldering the free hanging sample plate with a flat band copper wires to the stages base plate which is connected to the cold finger. The crystals itself are thermally contacted by gluing them the sample stage with carbon paint. That paint shows very good thermal conductivity without containing solvents that harm the organic crystals. The detailed construction sketches of the micro-stage are attached in the Appendix.

6.4. Transport measurements

Transport measurements describe the behavior of the system in the microwave range and the DC conductivity. In this frequency range the dielectric properties of the electronic system are probed. The DC response and the low frequency excitation to an applied electric field are given.

6.4.1. Four and two probe contact method for DC conductivity

Samples were contacted with a standard four-terminal method for in-plane transport measurements and two point contacts for measurements perpendicular to the plane. Contacts were made by evaporating 50Å gold pads on the samples, then 15 μm golden wires were glued with carbon paint to ensure electrical contact. In two-point measurements sometimes the wires are directly glued on the crystal. Vacuum grease on a sapphire holder ensures thermal contact. The contact geometry can be chosen in a way that the anisotropy of the samples can be measured. Temperatures from 300 K down to 2 K were recorded on a slow cooling rate of approximately 0.2-0.5 K per minute to avoid sample cracks. Currents are 50 μA at maximum to rule out electrical heating.

6.4.2. Microwave cavity perturbation

Microwave cavities are a tool for contact-less probing of low frequency transport properties [160–162]. They avoid problems in the injection contacts and are also able to bridge small cracks within the crystal. Small sample is placed into a cylindrical copper cavity which resonates in the TE_{011} mode. The shift of the center frequency and the change of its width with respect to an empty cavity are measured. Using cavity perturbation theory the response of the system can be calculated [22, 163–165]. Different cavities operated at 24 and 33.5 GHz were used. In plane and out of plane response of the samples are accessible by aligning the sample along or perpendicular the stacks with respect to the electric field within the cavity.

7. Data analysis

The measured reflectance data are merged and extrapolated depending on metallic or insulating behavior. Kramers Kronig relations are used to calculate the optical conductivity. For interpretation of the data a Drude Lorentz fit on the measured reflectivity and the extracted conductivity is performed. The extended Drude formalism is used to extract the frequency dependent scattering rate and the mass enhancement.

7.1. Cleaning and merging data

To extract the optical response all measured frequency ranges have to be merged into one reflectance spectrum. In case of IR microscopy data, the spectrum has to be cleaned from artificial features due to atmosphere. In the FIR range also beam splitter features occur. Further, loading effects of the broad range bolometer at low temperatures are to be compensated.

7.1.1. Atmospheric and beam-splitter artifacts

The atmospheric artifacts are due to the CO_2 and H_2O rotation and vibrational bands in the air which are IR active [37, 154]. They are located around 670 and 2350 cm^{-1} in case of CO_2 and around 1600 and 3700 cm^{-1} for H_2O. They are present since the IR microscope operates not under vacuum conditions. It is nitrogen flushed to reduce the atmospheric influence but the changes of the ambient conditions like temperature and humidity give small differences in the single energy spectra for the sample and reference measurement. In the case of cold finger cryostat (especially the micro cryostat) it could happen that due to non-perfect vacuum conditions small 'ice' artifacts show up as sharp lines in the range of the water lines on cooling. As shown in Fig. 7.1 the artifacts are cut and smoothed out in these cases. An additional line around 3000 cm^{-1} is due to an absorption feature in the KBr beam splitter.

In the case of beam-splitters, also the low frequency region is affected. The mylar FIR beam splitter shows an absorption around 60 and 120 cm^{-1}. In the case of the 50 μm mylar the spectrum cannot be trusted around them. Therefore 50 μm mylar measurements are just taken into account at lowest frequencies well below 50 cm^{-1}. Higher frequencies are just compared to check the overlap with the measurement of the upper frequency range. The 6 μm mylar beamsplitter is not affected by these absorption features.

7. Data analysis

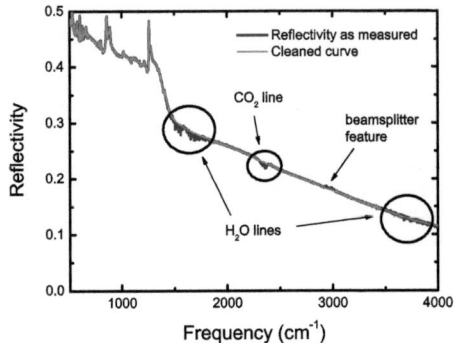

Figure 7.1.: Reflectivity spectrum of β''-(BEDT-TTF)$_2$SF$_5$CH$_2$CF$_2$SO$_3$ at 200 K measured with the IR microscope. The blue line shows the spectrum as measured and the orange one the spectrum cleaned from atmospheric and beam splitter artifacts.

7.1.2. Merging the data

The FTIR spectrometers and also the THz spectrometer cannot span the whole frequency range in one scan. Therefore several runs in different frequency ranges have to be performed. To merge all the data into one reflectance the results of the different ranges have to overlap perfectly to avoid steps or kinks. The overlap at high temperatures for all frequency ranges, from the submm setup THZ up to the VIS/UV range in the FTIR, is perfect. At temperatures below 70-100 K in the broad range FIR measurement loading effects in the bolometer appear. This measurement is performed with the 4.2 K bolometer and the 6 μm mylar beam splitter; optical filters are set in such a way that a frequency range up to 690 cm^{-1} down to 20-30 cm^{-1} can be measured. Due to this broad frequency range the detector feels the difference of the sample temperature to the environment as background radiation. This comes from the room temperature apertures and mirrors in the beam path. That leads to an additional loading of the sensitive element, which alters the sensitivity. The result is a reflectance curve, which is slightly off. The effect enhances the more room temperature mirrors are in between the sample and the detector and also with a higher temperature difference to the sample. Therefore the measured reflectivity value has to be corrected by 1-1.5% in the 30-100 K range and at lower temperatures by 1-2.5%. The loading effect is not present (or at least not detectable) in cases of the narrow bandwidth measurement. In the 1.4 K bolometer operated with the 50 μm mylar beamsplitter and optical filters, which block high frequencies, the range below the 60 cm^{-1} mylar feature is not affected and overlaps nicely to the submm measurements

7.2. Calculating the optical response and data fit

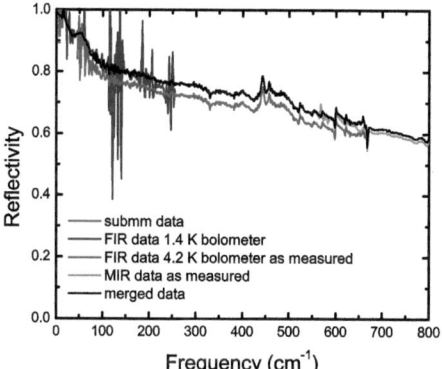

Figure 7.2.: Merged reflectivity spectrum (black) of β''-(BEDT-TTF)$_2$SF$_5$CH$_2$CF$_2$SO$_3$ at 100 K. Broad range FIR (red) and MIR (green) are scaled to fit. Around 40-60cm^{-1} additional narrow frequency measurements with the pumped bolometer and 6 μm mylar are taken (not shown for clarity). Additionally at lowest frequencies in the submm range and low frequency bolometer below 40 cm^{-1} a 20 point adjacent average smoothing is used.

in the THz range. Because of that the broad range bolometer data is scaled to fit the low frequency data. Then also MIR data is fitted better, however small scaling corrections of about 0.5% have to be done there as well. An example of merged data is given in Fig. 7.2 showing the loading effect of the broad band FIR measurement (red line).

7.2. Calculating the optical response and data fit

To obtain the optical response out of the measured reflectivity R, which is the amplitude of the complex reflectivity $r = Re^{i\phi}$, one has to make use of the Kramers Kronig relations. Writing the complex reflectivity as

$$\ln(r) = \ln(R) + i\phi \tag{7.1}$$

makes $\ln R$ and ϕ a Kramers Kronig consistent pair. Knowing $\ln(R)$ on the whole frequency range allows one to calculate the phase ϕ using the Kramers Kronig relation:

$$\phi(\omega) = -\frac{2\omega}{\pi} P \int_0^\infty \frac{\ln R(\omega')}{\omega' - \omega} d\omega', \tag{7.2}$$

7. Data analysis

with P the principal value of the integral. The revealed complex reflectivity is an optical response function of the system and can be converted into other response functions like the optical conductivity σ. Therefore the reflectance in the whole frequency range has to be known. The data has to be extrapolated sufficiently to low and high frequencies.
At high frequencies, above the reflectance minimum around 6000 cm^{-1}, no temperature dependence was found. As extrapolation the room temperature reflectance data in the VIS range is merged to all temperatures to account for the interband transitions. Above the highest measured frequencies an ω^{-2} drop till 100000 cm^{-1} and an ω^{-4} drop till 250000 cm^{-1} is extrapolated to describe the transparent regime, c.f. Sec 2.1.1. In the case of measurements along the insulating direction, the reflectivity was extrapolated flat up to 30000 cm^{-1} and then the ω^{-2} and ω^{-4} drops follow.
At low frequencies the right extrapolation depends a lot on the systems behavior. For perfect metals the reflectivity is expected to go to 1 at zero frequency as described by the Hagen Rubens formula $1 - \alpha\omega^{-2}$ Eqn. (2.3). The metallic systems (judged from DC) are extrapolated using the pre-factor α such that the extrapolation fits with the lowest measured reflectivity point. In the case of insulating behaviors the reflectivities are just extrapolated with a straight line of constant reflectivity.

7.2.1. Drude Lorentz fit

For a first interpretation of the optical response, the measured reflectivity and the extracted conductivity are fitted simultaneously based on the Drude Lorentz model (Sec. 2.1.1). The zero frequency peak is modeled by a Drude while the bands are modeled with broad Lorentz curves. Sharp Lorentzians are used to describe uncoupled phonons while the ones coupled to the electronic background are described by a Fano lineshape (Sec. 3.3.2).

7.2.2. Extended Drude formalism

The extended Drude model (Sec. 2.1.2) can describe the low frequency response introducing a frequency dependent scattering rate. Here it is used to trace the interactions of the charge carriers with a collective mode. The plasma frequency entering the extended Drude formula usually is taken from the integrated spectral weight. Here the only the onset to the zero frequency response could be measured. Therefore the plasma frequency is taken from the Drude Lorentz fit of the reflectance and conductivity data.

Part III.

Results

8. Transport measurements

The transport properties of the two dimensional organic conductors and superconductors are discussed. DC and microwave techniques at 24 and 33.5 GHz probe the β''-(BEDT-TTF)$_2$SF$_5$CH$_2$CF$_2$SO$_3$ superconductor and its deuterated compound as well as the isostructural metal β''-(BEDT-TTF)$_2$-SF$_5$CHFSO$_3$. The superconductors basically show metallic behavior down to the superconducting transition at $T_c = 5$ K. Deviation from a simple Drude behavior point out its proximity to charge order. Also the metallic sister compound has some indications for localization in the out-of-plane direction. Two kind of samples of the superconductor β-(EDT-TTF)$_4$[Hg$_3$I$_8$]$_{(1-x)}$ are characterized using DC measurements. They are metallic down to the superconducting transition around $T_c = 8$ K. However, some samples do not become superconducting and turn insulating instead.
α-(BEDT-TTF)$_2$I$_3$ is an example for a charge ordered insulating state. At 135 K a strong metal to insulator transition takes place. The resistivity increases orders of magnitude.

8.1. The organic superconductor β''-(BEDT-TTF)$_2$SF$_5$CH$_2$CF$_2$SO$_3$

The DC measurements of the β''-(BEDT-TTF)$_2$SF$_5$CH$_2$CF$_2$SO$_3$ superconductor for the in- and out-of-plane properties are shown in Fig. 8.1 (a) in the upper and lower graph respectively. The microwave behavior in-plane at 24 and 33.5 GHz is given in Fig. 8.1 (b). Both, the DC and microwave properties, are metallic when cooled down from ambient temperature. The behaviors differ right before entering the superconducting transition around $T_c \approx 5$ K as shown in the respective insets. The DC properties stay metallic, but the microwave properties show first a resistivity increase before turning superconducting. This increase is a first trace to charge fluctuation in the system.
In the DC measurements of β''-(BEDT-TTF)$_2$SF$_5$CH$_2$CF$_2$SO$_3$ (Fig. 8.1) the absolute value at room temperature is $\rho_a = 2.6$ Ωcm. That is in agreement with previous reports [66, 115]. The in-plane out-of-plane anisotropy at room temperature is about $\rho_c/\rho_a \approx 140$. X-band ESR measurements by Wang et al. [166] estimate an in-plane anisotropy of $\sigma_{max}/\sigma_{min} = 1.35$. This evidences the pronounced two-dimensionality of the system which is verified in the optical measurements (Sec.10). The resistivity ratio of high to low temperatures $R(300\text{ K})/R(6\text{ K})$ is approximately 50 for the in-plane and 150 for the interlayer transport. At low temperatures there is a linear behavior with temperature up to 125 K for both directions. Refs. [121, 132] report a T^2 dependence below 10 K which

8. Transport measurements

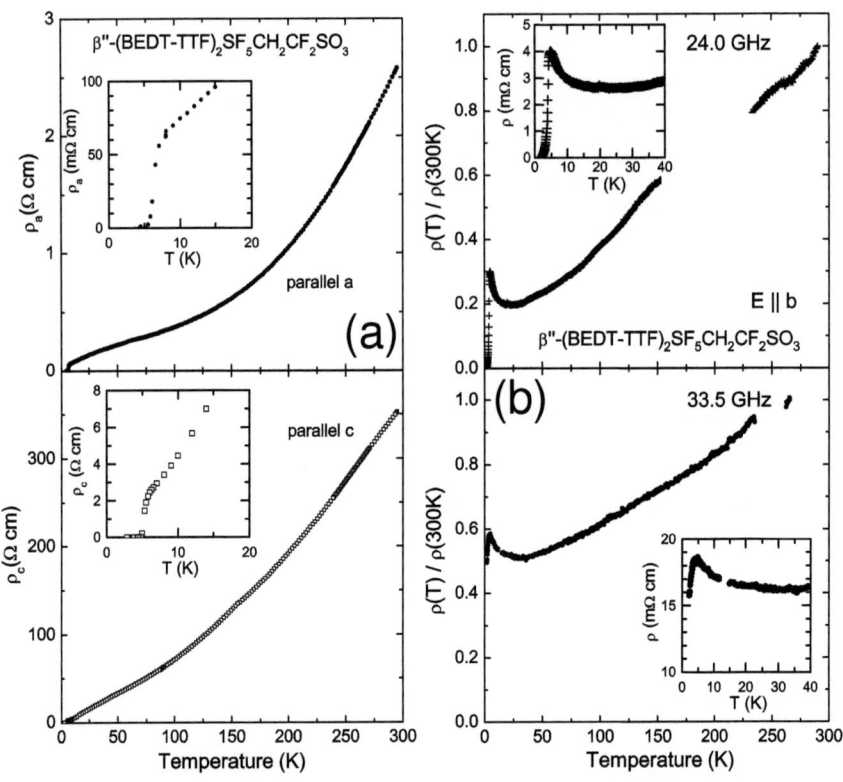

Figure 8.1.: Transport properties of the β''-(BEDT-TTF)$_2$SF$_5$CH$_2$CF$_2$SO$_3$ organic superconductor. (a) In- and out of plane dc-resistivity. (b) In-plane microwave resistivity at 24 and 33.5 GHz.

8.1. The organic superconductor β''-(BEDT-TTF)$_2$SF$_5$CH$_2$CF$_2$SO$_3$

turns into a linear behavior above. However, pressure dependent studies [121] oppose its origin in electron-electron scattering (also c.f. overview in Sec. 4.1.1). We assign the linear slope with the onset of charge fluctuations in the system which alter the metallic state. It was theoretically predicted [18, 167] and experimentally observed in α-salts [78] that a T^2 dependence of the scattering rate changes into a linear behavior on the influence of charge fluctuations. Another slope change at 200 K is along the interlayer direction $\rho_c(T)$. A direct proof of this being the onset of charge-order patterns around 200-150 K is given by vibrational spectroscopy in the subsequent Sec. 9. The critical temperature for the superconducting transition is determined from the drop in resistivity along both directions. It sets in at T_c =5.9 K. The width of the transition is $\Delta T_c \approx$ 1 K. The high onset of superconductivity compared to literature [83, 115] evidences the excellent quality of the single crystals.

The microwave properties of β''-(BEDT-TTF)$_2$SF$_5$CH$_2$CF$_2$SO$_3$ plotted in Fig. 8.1, are normalized to the room temperature values. That is because uncertainties in the absolute values can exceed a factor of 10. This is due to errors in the depolarization factor caused by the irregular shape of the samples. However, an estimation of the microwave resistivity leads to about $\rho(300$ K$)\approx 30$ mΩcm, one order of magnitude less compared to the dc values. The microwave properties at 24 and 33.5 GHz are metallic down to 20-30 K. On further cooling the resistivity increases by 20 to 30 % before becoming superconducting as shown in the insets. The superconducting transitions are at about 5 K for the 24 GHz and around 4.5 K for the 33.5 GHz measurement. This increase of resistivity is similar to what is known for dc properties of several other systems [17, 120, 168–172]. Therefore we attribute this behavior to charge fluctuations within the system.

8. Transport measurements

8.2. The deuterated superconductor β''-(d$_8$-BEDT-TTF)$_2$SF$_5$CH$_2$CF$_2$SO$_3$

For the deuterated sister compound β''-(d$_8$-BEDT-TTF)$_2$SF$_5$CH$_2$CF$_2$SO$_3$ dc and microwave properties are also metallic at low temperatures (Fig. 8.2(a) and (b)) before it turns superconducting. DC properties for the out of plane direction exhibit a hump on cooling from room temperature. Then the behavior turns metallic.
In DC (Fig. 8.2(a)) the resistivity ratio from high to low temperatures is $R(300$ K$)/R(6$ K$)=100$. The low temperature power law $\rho(T) \propto T^\alpha$ turns out to be slightly lower than quadratic. The transition into the superconducting state sets in at $T_c = 5.6$ K with a width of approximately 1 K. In the c-direction perpendicular to the conducting plane the resistivity increases with decreasing temperature until a maximum is reached around 200 K. A nearly linear drop in resistivity below 150 K follows on further cooling. Again there is a slight slope change around 100 K and a small kink at 20 K.
The microwave resistivity (Fig. 8.2(b)) at room temperature is about 1 Ωcm, comparable to the DC values. The resistivity ratios are the same for both frequencies ($\rho(300$ K$)/\rho(6$ K$)\approx10$). The behavior at 33.5 GHz is nearly quadratic as indicated by the fit while at 24 GHz there is a big drop at 200 K. Below 50 K both measurements have only a weak temperature dependence. Around 5.6 K superconductivity sets in.

8.3. The organic metal β''-(BEDT-TTF)$_2$SF$_5$CHFSO$_3$

The organic metal β''-(BEDT-TTF)$_2$SF$_5$CHFSO$_3$ is the isostructural sister compound to the superconductors. Its DC and microwave properties are plotted in Fig. 8.3. Microwave and in plane properties are metallic while the out-of-plane resistivity shows localization effects at low temperatures. Also some of the investigated crystals exhibit a hump in resistivity before turning metallic for low temperatures.
DC properties (Fig. 8.3 (a)) are metallic down to 50 K in the in-plane as well as in the out-of-plane direction. Below 50 K the resistivity drop stops and becomes nearly temperature independent for the in-plane direction. In the interplane directions it turns insulating. The room temperature value 2.6 Ωcm is comparable to the superconductor. The high to low temperature ratio is about $\rho_b(300$ K$)/\rho_b(30$ K$)=15.3$. For the in-plane out-of-plane anisotropy one finds $\rho_c/\rho_b \approx 200$. In the out of plane direction an insulating behavior sets in below 50 K. The resistivity rapidly increases even above the room temperature values.
Some of the samples (two out of eight) have a different temperature dependence of the resistivity as shown in Fig. 8.3 (b). Starting at room temperature from equivalent absolute values as in the pure metallic sample, the resistivity slightly increases on cooling to 250 K. Then the behavior turns semiconducting with the resistivity growing by a factor of 4 on cooling to 70 K. There it exhibits a broad maximum. To low temperatures the in-plane resistivity drops with a kink at 16 K. A similar behavior for that compound is reported in literature [83].
To check the semiconducting or insulating behavior in the resistivity curves ESR mea-

8.3. The organic metal β''-$(BEDT\text{-}TTF)_2SF_5CHFSO_3$

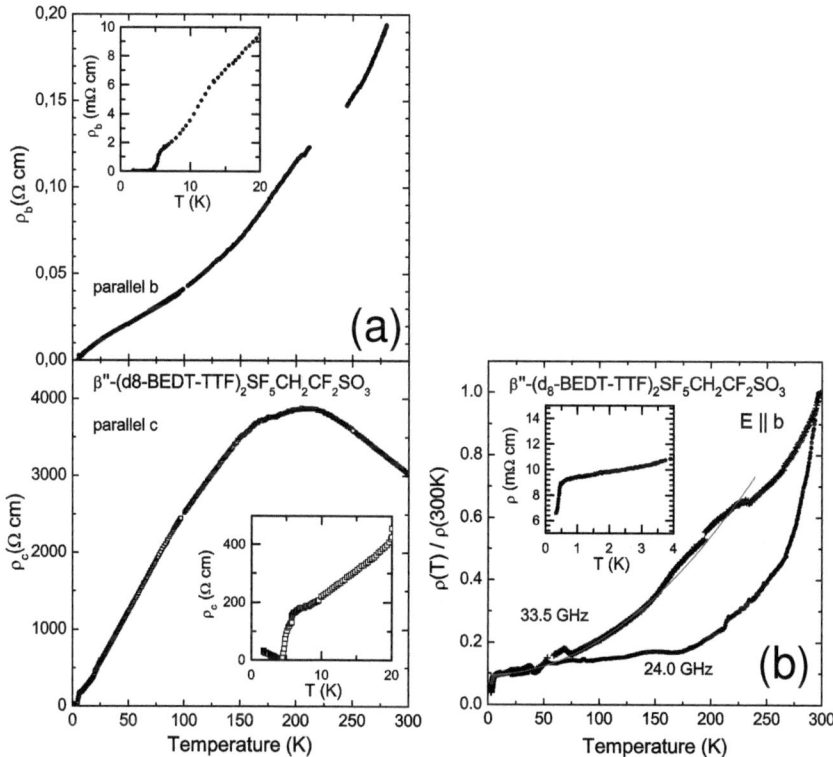

Figure 8.2.: Transport properties of the deuterated β''-$(d_8\text{-}BEDT\text{-}TTF)_2SF_5CH_2CF_2SO_3$ organic superconductor. (a) In- and out of plane dc-resistivity. (b) In-plane microwave resistivity at 24 and 33.5 GHz.

8. Transport measurements

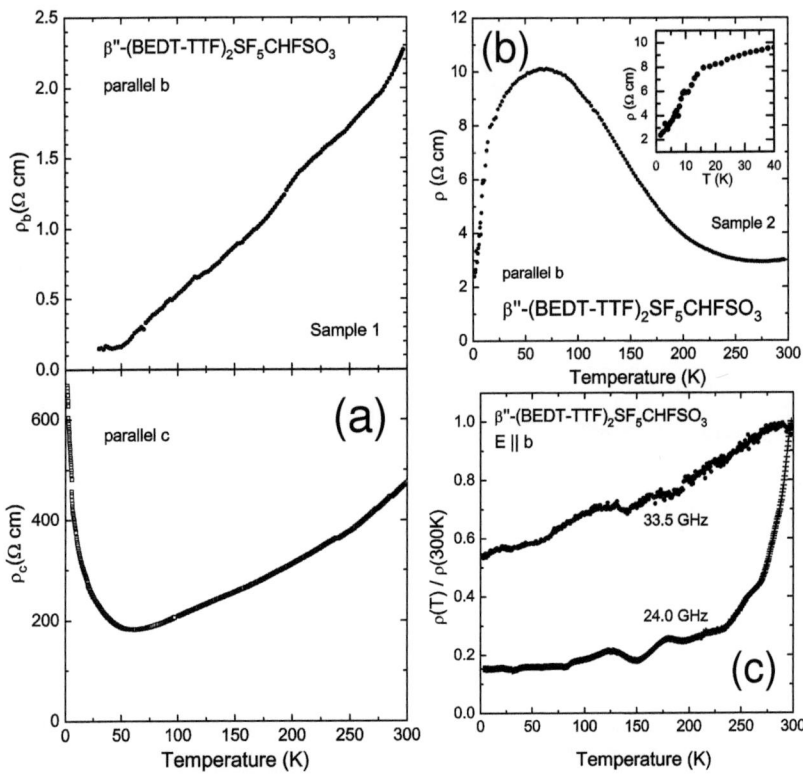

Figure 8.3.: Transport properties of the β''-(BEDT-TTF)$_2$SF$_5$CHFSO$_3$ organic metal. (a) In- and out of plane dc-resistivity of sample 1. (b) In-plane resistivity of sample 2. (c) In-plane microwave resistivity at 24 and 33.5 GHz.

8.3. The organic metal β''-(BEDT-TTF)$_2$SF$_5$CHFSO$_3$

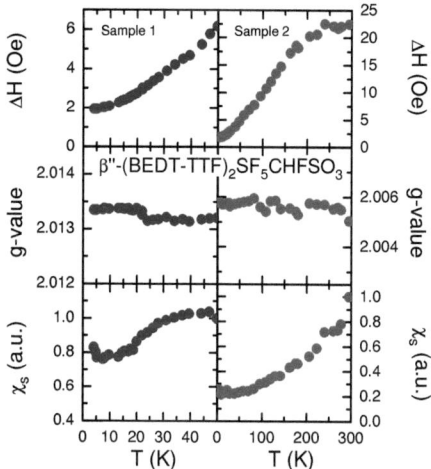

Figure 8.4.: ESR properties of two different β''-(BEDT-TTF)$_2$SF$_5$CHFSO$_3$ samples. Sample 1 shows pure metallic behavior in the in-plane dc properties and sample 2 exhibits a resistivity hump around 70 K. From S. Yasin at 1. Physikalisches Institut Universität Stuttgart.

surements are performed on 2 samples of the organic metal. Sample 1 shows a metallic in-plane behavior and sample 2 a hump around 70 K. As shown in Fig. 8.4 there are no significant differences in the ESR parameters between the two samples. There is no hint for magnetic ordering effects. The linewidth of $\Delta H = 25$ Oe and the g values of 2.013 and 2.003 at room temperature are comparable to the results of Ref. [83]. Like there, the linewidth drops to about 2 Oe on cooling to lowest temperature. The spin susceptibility in sample 1 does not show strong dynamics on cooling. Below 50 K the small drop suggests a slightly less metallic behavior and corresponds to the flattening of the resistivity curve [Fig. 8.3 (a)] at lowest temperatures. The constant spin susceptibility below 20 K indicates pure metallic behavior. In sample 2 a gradual drop in spin susceptibility is found. Below 100 K it stays nearly constant which points to a metallic behavior opposing the trend of the DC resistivity [Fig. 8.3 (b)].

Also the microwave properties [Fig. 8.3 (c)] are purely metallic down to lowest temperatures. On cooling for 24 GHz the resistivity has a steep drop and then only little variation below 150 K. Compared to that, the 33.5 GHz measurement reveals a gradual and almost linear drop in resistivity over the whole temperature range.

8. Transport measurements

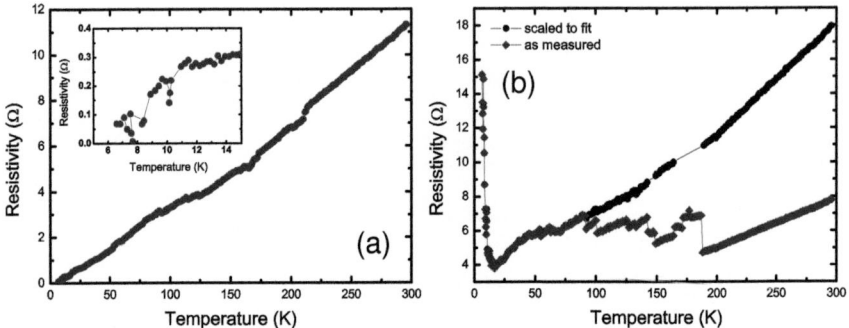

Figure 8.5.: DC conductivity of the β-(EDT-TTF)$_4$[Hg$_3$I$_8$]$_{(1-x)}$. (a) A sample staying metallic down to the superconducting transition region around $T_c \approx 8$ K. This sample was used in vibrational spectroscopy. (b) A sample turning insulating at lowest temperatures instead of turning superconducting. A lot of cracks occurred on cooling, the high temperature data was scaled to fit the low temperatures. This sample was used at the in-plane optical spectroscopy.

8.4. The EDT based superconductor β-(EDT-TTF)$_4$[Hg$_3$I$_8$]$_{(1-x)}$

The β-(EDT-TTF)$_4$[Hg$_3$I$_8$]$_{(1-x)}$ is synthesized with the goal of a superconducting crystal. However, as introduced in Sec. 4.2 even the crystals from the same sample batch do not always become superconducting. Some turn insulating at low temperatures due to slightly different stoichiometry disturbance in the anion chain. Applying a small pressure of about 300 bar makes them superconducting. Among the samples investigated both kind of samples are found. The sample labeled sample A stays metallic and turns superconducting. Its dc properties are given in Fig. 8.5 (a). The resistivity drops linear with decreasing temperature. There are only slight deviations in the range from 100-220 K. The inset shows the region around the superconducting transition temperature. Below 12 K the drops has a significant steeper slope down to ≈ 9 K. The crystal becomes superconducting around $T_c \approx 8$ K while the transition $\Delta T \approx 2$ K is rather broad. This indicates a low quality of the crystal compared to the β''-(BEDT-TTF)$_2$-SF$_5$CH$_2$CF$_2$SO$_3$. This might be due to a lot of cracks within the system.

The other sample, labeled sample B, behaves metallic on cooling from room temperature. But instead of becoming superconducting it turns insulating as shown in Fig. 8.5 (b). Significant stress and strain is believed to turn the sample insulating instead of superconducting [134]. A hint for that could be the cracks that occur on cooling. They are seen in the measurement as steps in the curve. They are labeled 'as measured' in the graph. The high temperature data is therefore scaled to fit the low temperatures. Again, there is a linear drop in resistivity with decreasing temperature down to 200 K then the slope changes. Below it is difficult to judge due to the cracks. Scaling the data to account for the cracks reveals a linear drop of resistivity down to 70 K before the slope flattens again. Below 50 K the resistivity drops down again. At 17 K it turns insulating down to lowest temperatures. The shown behaviors, the superconducting and insulating one, are in perfect agreement with characterizations found in literature [134].

The fact that crystals from the same batch can show either show superconducting or insulating behavior already indicates how close charge-ordering effects are related to superconductivity. Smallest changes in the anion configuration change the chemical pressure and therefore the effective electronic correlations. But also the huge influence of cracks in the insulating and the broad transition in the superconducting sample show how fragile these systems are to impurities and imperfections. These are visible in the optical pre-characterization to find the optical axes which revealed a lot of twinned crystals in the batches. Nevertheless, the surface quality of the crystals was excellent. Finally on sample A (superconducting) it was possible to perform the vibrational spectroscopy while on sample B (insulating) for the first time a temperature dependent in-plane optical spectrum could be measured.

8.5. The metal insulator transition in α-(BEDT-TTF)$_2$I$_3$

To investigate the metal insulator transition in systems with pronounced charge order, the dc transport in α-(BEDT-TTF)$_2$I$_3$ is measured. Its resistivity curve is given in Fig. 8.6. Starting from high temperatures, the system behaves metallic on cooling. At 135 K the metal insulator transition takes place. The resistivity increase exceeds two orders of magnitude. On further cooling the resistivity further increases the total increase spanning more then four orders of magnitude by cooling to 100 K and going on to increase to lower temperatures. That indicates that the charge order is very strong and stable. Basically no charge carriers are left to contribute to a coherent transport. In comparison to the former hints for localization in the metallic and superconducting compounds, it is obvious that the resistivity in a real charge ordering metal-insulator transition increases orders of magnitudes more.

8.6. Summary of transport properties

In summary the different systems can be successfully characterized by their transport properties showing metallic, insulating, or superconducting behavior. At high temperatures all systems basically do not show a resistivity saturation in their in-plane properties. The β''-(BEDT-TTF)$_2$SF$_5$CHFSO$_3$ which is expected to be the weakest correlated system shows a metallic behavior down to lowest temperatures. However in the out-of-plane direction an upturn in resistivity was found to lowest temperatures. Some samples showed an activated behavior even in the in-plane direction before turning metallic at low temperatures. Microwave properties on the other hand reveal pure metallic behavior and also ESR measurements did not show any traces for ordering effects.

The systems with high electron-electron correlations, here the α-(BEDT-TTF)$_2$I$_3$ are

8. Transport measurements

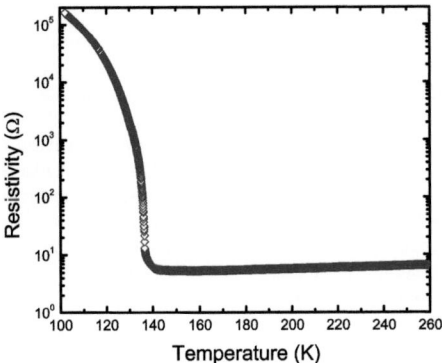

Figure 8.6.: In-plane resistivity vs temperature for α-(BEDT-TTF)$_2$I$_3$ from room temperature down to the metal-insulator transition.

expected to undergo a metal-insulator transition at the charge ordering temperature. Indeed the samples show a strong and abrupt resistivity increase at the phase transition exceeding orders of magnitude.

The superconducting systems, the β''-(BEDT-TTF)$_2$SF$_5$CH$_2$CF$_2$SO$_3$, its deuterated sister compound β''-(d$_8$-BEDT-TTF)$_2$SF$_5$CH$_2$CF$_2$SO$_3$, and the β-(EDT-TTF)$_4$[Hg$_3$I$_8$]$_{(1-x)}$ showed metallic transport properties down to the superconducting transition. However, they all show traces of charge order or charge fluctuation by respective slope changes in the temperature dependent resistivity. The β''-(BEDT-TTF)$_2$SF$_5$CH$_2$CF$_2$SO$_3$ shows even a small hump in its microwave resistivity before turning superconducting. For the β-(EDT-TTF)$_4$[Hg$_3$I$_8$]$_{(1-x)}$ system some samples even within one batch turned either superconducting or insulating at lowest temperatures. These behaviors suggest the superconducting systems to be closely related to charge order and influenced by charge fluctuations in the vicinity to the charge order transition.

9. Vibrational spectroscopy

The vibrational properties of the organic conductors and superconductors provide information on the amount of charge per molecular site, possible structural changes, as well as insight into the coupling of the vibrational system to the electronic background via e-mv coupling. The superconducting systems show a small charge disproportionation of about $0.2e$ while at the same time they show metallic transport properties (Sec. 8). This proves the presence of a charge-ordered metallic state. Systems with a metal-insulator transition show a charge disproportionation well above $0.5e$ in the insulating state. In the metallic compounds no or only weak traces of charge redistributions are found in the IR active modes. Slight charge differences of about $0.1e$ can be traced even in the metals. In the case of α-$(BEDT$-$TTF)_2I_3$ the possibility of an additional structural phase transition in the system is investigated in comparison to θ-$(BEDT$-$TTF)_2RbZn(SCN)_4$.

9.1. On-site charge distribution in the organic superconductors and metals

Main focus of the vibrational measurements is the investigation of the charge redistribution in the β''-$(BEDT$-$TTF)_2SF_5CH_2CF_2SO_3$ organic superconductor and its metallic sister compound β''-$(BEDT$-$TTF)_2SF_5CHFSO_3$ using IR and Raman spectroscopy. As described in Sec. 3.3.1, the frequency of the C=C double bond modes directly probe the charge on the molecule. Further, the organic superconductor β-$(EDT$-$TTF)_4[Hg_3I_8]_{(1-x)}$ the family of the organic superconductor α-$(BEDT$-$TTF)_2NH_4Hg(SCN)_4$ and the corresponding strongly renormalized metal α-$(BEDT$-$TTF)_2TlHg(SCN)_4$ are investigated with IR vibrational spectroscopy. The existence of charge disproportionation in the superconductors should be proven while in the metals less or no traces for charge order are expected.

9.1.1. The organic superconductor β''-$(BEDT$-$TTF)_2SF_5CH_2CF_2SO_3$

IR vibrational spectroscopy

As described in Sec. 6.3.3 the vibrational spectra of the systems are measured with the light polarized along the insulating crystal direction perpendicular to the conducting layer to observe the charge sensitive $B_{1u}(\nu_{27})$ vibrational mode which is absent in other polarizations. Also the measurement along the insulating direction avoids screening of the

9. Vibrational spectroscopy

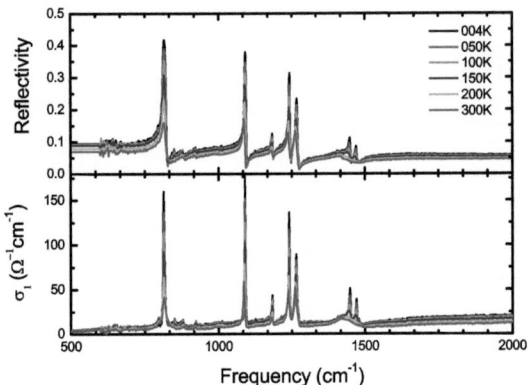

Figure 9.1.: Reflectivity and conductivity of β''-(BEDT-TTF)$_2$SF$_5$CH$_2$CF$_2$SO$_3$ along its insulating c-direction. Temperature evolution and hardening of the BEDT-TTF molecular vibrations and the vibrational modes of the SF$_5$CH$_2$CF$_2$SO$_3$ anion.

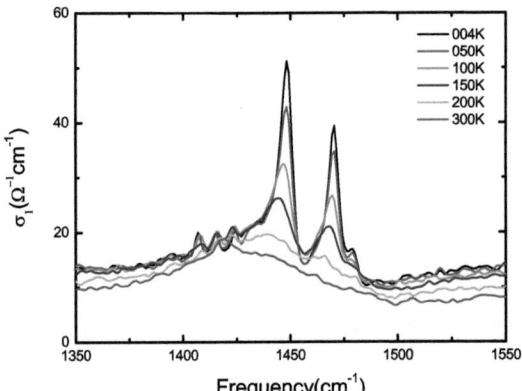

Figure 9.2.: Splitting of the charge sensitive $B_{1u}(\nu_{27})$ molecular vibration in the conductivity spectra of β''-(BEDT-TTF)$_2$SF$_5$CH$_2$CF$_2$SO$_3$ with decreasing temperature. The optical conductivity is probed with polarized light perpendicular to the conducting layers.

9.1. On-site charge distribution in the organic superconductors and metals

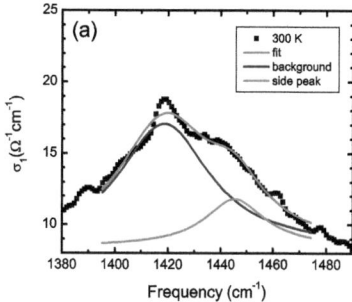
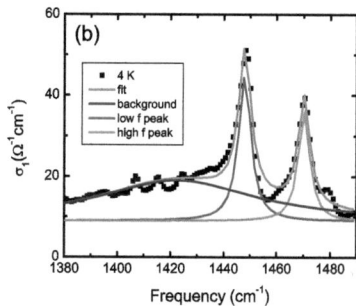

Figure 9.3.: Fit of the contributions to the $B_{1u}(\nu_{27})$ mode in β''-(BEDT-TTF)$_2$-SF$_5$CH$_2$CF$_2$SO$_3$ at (a) 300 K and (b) 4 K.

modes by the electronic background. The temperature dependent reflectivity and conductivity spectra of β''-(BEDT-TTF)$_2$SF$_5$CH$_2$CF$_2$SO$_3$ along this direction are shown in Fig. 9.1. The SF$_5$CH$_2$CF$_2$SO$_3$ anion is rather large, complex, and of low symmetry. Therefore it exhibits a lot of IR active modes. Most of the observed modes are due to anion vibrations. The vibrational modes of the BEDT-TTF molecule are mainly symmetric and therefore not IR active. An assignment of the modes can be done based on literature [66, 93, 97, 99]. With decreasing temperature the modes sharpen and harden a few wavenumbers due to thermal contraction of the molecular bonds. The anion modes do not show any sign of splitting or changes in frequency position beyond this hardening. No additional modes appear. Also the non charge sensitive modes of the BEDT-TTF molecule do not change. That proves that there is no structural change that causes a symmetry change.

The most interesting and significant change happens at the charge sensitive $B_{1u}(\nu_{27})$ C=C double bond vibration located around 1450 cm^{-1} (Fig. 9.2). At room temperature the broad band of this vibration is located at 1423 cm^{-1}. The mode at 1418 cm^{-1} is identified as CH$_2$ bending mode ν_{28}. However it is also possible to fit a side peak into the broad mode at room temperature (Fig. 9.3 (a)). In that case there is a broad contribution around 1418 cm^{-1} (blue line) and a small side peak at 1445 cm^{-1} (green line). The spectral weight distribution between these two peaks can be modeled in several ways. Also the side peak can be assumed as $B_{1u}(\nu_{27})$ mode and the low frequency peak stays as background.

With decreasing temperature the broad $B_{1u}(\nu_{27})$ mode hardens and splits. The hardening is strongest in the temperature range down to 150-100 K since in that range the largest thermal shrinking of the system takes place. The onset of a splitting starts al-

9. Vibrational spectroscopy

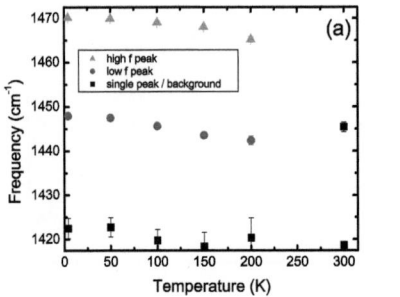 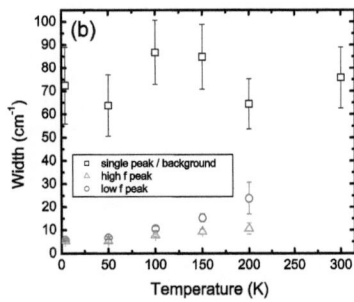

Figure 9.4.: Temperature dependence of the (a) position and (b) width of the high (green) and low (red) frequency peak to the split $B_{1u}(\nu_{27})$ mode in β''-(BEDT-TTF)$_2$-SF$_5$CH$_2$CF$_2$SO$_3$ showing the charge redistribution. The broad and nearly temperature independent background peak is given by the black squares. At 300 K possibly there is a side peak in the background peak that can be assigned as the non-split $B_{1u}(\nu_{27})$ mode.

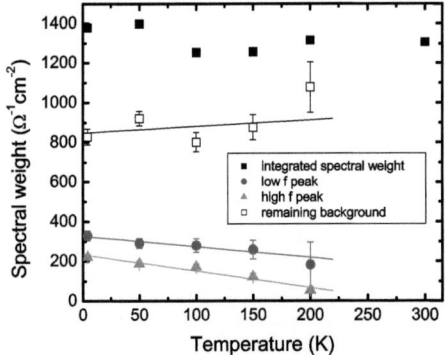

Figure 9.5.: Evolution of spectral weight of the contributions to the $B_{1u}(\nu_{27})$ mode in β''-(BEDT-TTF)$_2$SF$_5$CH$_2$CF$_2$SO$_3$. The split mode enhances on expense of the remaining background. The background is calculated by subtracting the two peak contributions from the integrated spectral weight. See text.

9.1. On-site charge distribution in the organic superconductors and metals

ready at 200 K; it becomes well developed below 150 K. That is exactly the temperature range in which the DC properties show a slope change in the resistivity. At the low temperatures the mode splits into two narrow contributions (red and green) as shown in Fig. 9.3 (b) for 4 K. At all temperatures a broad background contribution (blue) around 1418 cm^{-1} stays present. In a fit it is not possible to distinguish between this background peak and an eventually remaining side peak as in the room temperature spectrum. The temperature dependence of position and width of the fitted contributions is given in Fig. 9.4 (a) and (b). It visualizes again the splitting of the single mode (the upper black square at 300 K is for the high frequency side peak) at room temperature into a split mode (red circles, green triangulars) below 200 K including the hardening and narrowing with decreasing temperature. The splitting (difference frequency) between the modes stays basically constant for all temperatures below the splitting temperature. The background peak (black squares) is nearly temperature independent. It hardens only slightly while its width stays constant within the errorbar.

The integrated spectral weight of the $B_{1u}(\nu_{27})$ band (including the background peak) is conserved on cooling as shown as solid black squares in Fig. 9.5. The spectral weight redistribution with decreasing temperature shows that the two contributions of the split mode increase on expense of the broad mode. The low (red) and high (green) frequency peaks of the $B_{1u}(\nu_{27})$ mode increase linearly with decreasing temperature while the background decreases. It is not possible to distinguish between a background contribution and a possible side peak as at room temperature. Therefore both are fitted combined using a single peak. Its decrease in spectral weight on cooling is then calculated from the total integrated spectral weight reduced by the spectral weight of the split mode. The result is given in Fig. 9.5 as open squares. Assuming the background as temperature independent that would mean that the reduction of the remaining background spectral weight is due to the reduced side peak spectral weight.

This behavior clearly shows the onset of a charge redistribution between the molecules below 200 K. The broad room temperature mode corresponds to an average charge of +0.5e per molecular site as expected for a quarter-filled metallic system. The broad peak evidences a large thermal broadening of the smeared-out charge distribution or fluctuation of the conducting charge carriers. On cooling the charge carriers start to form local charge order patterns, seen by the appearance of the splitting and sharpening of the modes. The frequency splitting of $\Delta\nu$ =22 cm^{-1} is related to a charge disproportionation between the sites of approximately 0.21e assuming a slope of 105 cm^{-1}/e [70, 100, 105–107] (Fig. 3.11). With metallic transport properties all the way down to the superconducting transition, this result clearly proves the existence of a charge-ordered metallic state. The charge-order fluctuations lead to a time-averaged disproportionation of charges between the molecules evidenced here by the splitting of the charge sensitive $B_{1u}(\nu_{27})$ mode. Since the splitting between the peaks stays constant on cooling there is no change in the degree of charge redistribution.

9. Vibrational spectroscopy

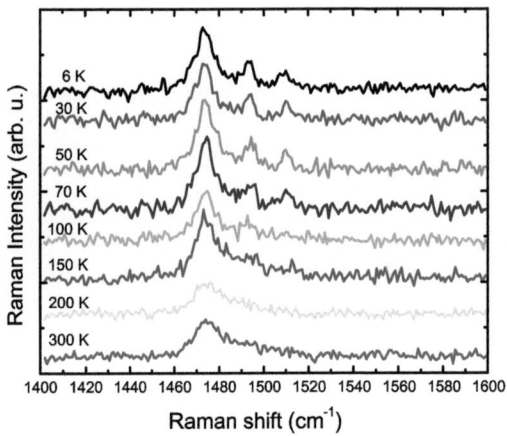

Figure 9.6.: Temperature dependence of the charge sensitive split $A_g(\nu_2)$ and $A_g(\nu_3)$ C=C double bond modes in β''-(BEDT-TTF)$_2$SF$_5$CH$_2$CF$_2$SO$_3$. Due to different acquisition times during the measurement all lines are normalized and shifted against each other for clarity.

Raman spectroscopy

Raman spectroscopy probes the fully symmetric modes which are not active in the IR. Of interest here are the charge sensitive $A_g(\nu_2)$ and $A_g(\nu_3)$ vibrations to probe the average charge per molecular site. The temperature dependent Raman spectra (normalized to account for different acquisition times and shifted with respect to each other for clarity) are shown in Fig. 9.6. At room temperature only an asymmetric peak at 1474 cm^{-1} is present and assigned to the $A_g(\nu_3)$ vibration. However it is also possible to fit a second peak at 1492 cm^{-1} to the high frequency shoulder of the peak as shown in Fig. 9.7 (a). The position of the $A_g(\nu_3)$ mode corresponds to an average site charge of 0.5e [100]. This is in agreement with the results from the IR measurement with the non split charge sensitive $B_{1u}(\nu_{27})$ mode at room temperature. The mode around 1492 cm^{-1} could be assigned to the $A_g(\nu_2)$.

With the temperature decreasing below 200 K clearly two additional modes at 1493 cm^{-1} and 1510 cm^{-1} appear besides the peak at 1474 cm^{-1}. They can be well distinguished below 150 K. A fit to all 3 modes at 6 K is shown in Fig. 9.7 (b). Their temperature dependence is given in Fig. 9.8 for the Raman shift and in Fig. 9.9 for the width of

9.1. On-site charge distribution in the organic superconductors and metals

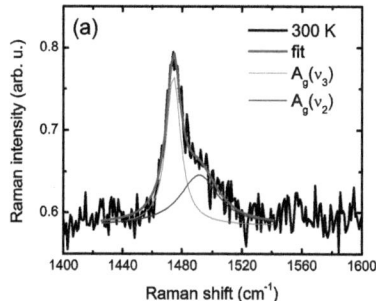 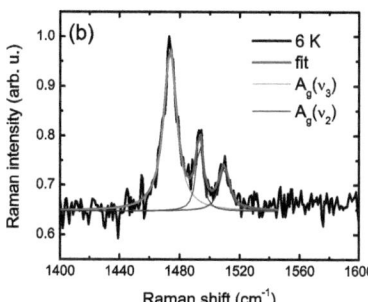

Figure 9.7.: Fit of the contributions to the $A_g(\nu_3)$ and $A_g(\nu_2)$ mode in β''-(BEDT-TTF)$_2$-SF$_5$CH$_2$CF$_2$SO$_3$ at (a) 300 K and (b) 6 K.

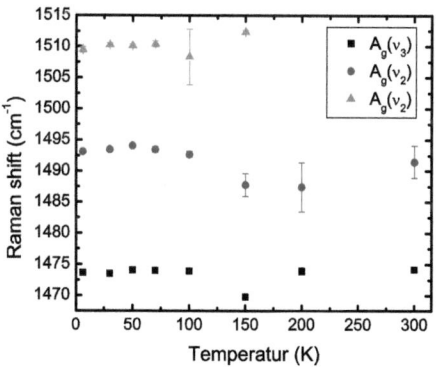

Figure 9.8.: Raman shift of the $A_g(\nu_2)$ and $A_g(\nu_3)$ modes in β''-(BEDT-TTF)$_2$-SF$_5$CH$_2$CF$_2$SO$_3$.

123

9. Vibrational spectroscopy

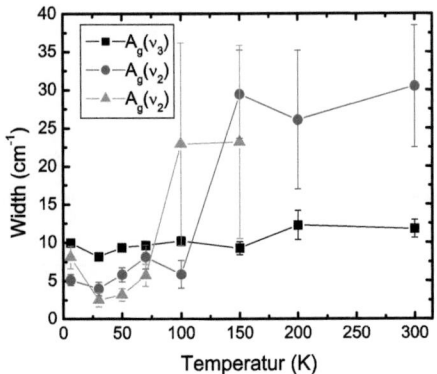

Figure 9.9.: Width of the $A_g(\nu_2)$ and $A_g(\nu_3)$ modes in β''-(BEDT-TTF)$_2$SF$_5$CH$_2$CF$_2$SO$_3$.

the modes. The mode at 1474 cm^{-1}, identified with the $A_g(\nu_3)$, shows no temperature dependence in its Raman shift (black squares). Just a slight red shift at 150 K, right where transport and IR vibrational spectroscopy saw the onset of charge fluctuations. At lower temperatures it shifts directly back, again. Also the width of that mode decreases slightly at 150 K and stays reduced. In addition, right at that temperature two additional peaks become visible (red circle, green triangular). However, also above 150 K the low-frequency peak (red circles) is already observed as the side peak under the $A_g(\nu_3)$ mode. This peak red shifts starting from room temperature down to 150 K. Below that temperature it shifts back and shows a temperature independent behavior from 100 K down to lowest temperatures. The high frequency peak (green triangles) slightly red-shifts on cooling from 150 to 100 K. For lower temperatures its Raman shift basically stays constant. The width of both peaks narrow from about 25 cm^{-1} to 5 cm^{-1} in the range from 100-150 K. Then their width stays narrow. In addition, their relative peak strength are both of comparable peak size (data normalized to $A_g(\nu_3)$). Therefore, and with respect to the equivalent mode in the metallic sister compound (discussed later), we assign the two peaks to the $A_g(\nu_2)$ mode which is split due to the charge redistribution below 150 K.

The frequency difference of $\Delta\nu = 17$ cm^{-1} between the contributions corresponds to a charge disproportionation between the sites of 0.2e [100] in perfect agreement with the value concluded from the $B_{1u}(\nu_{27})$ splitting in the IR. The constant splitting between the peaks shows that no change in the charge disproportionation takes place on cooling and that the charge redistribution sets in abrupt. However, it is unclear why the non-split

$A_g(\nu_2)$ mode above 150 K is found at too low frequency.

One would also expect the $A_g(\nu_3)$ mode to split due to the charge disproportionation. But the experimental fact that the $A_g(\nu_3)$ vibration is typically highly e-mv coupled suppresses the splitting as suggested by Ref. [74]. In addition, its strong asymmetric shape is a hint for the presence of charge fluctuations because the mode is highly coupled to the charge-transfer excitation [173]. The redistributed charge is basically smeared out with the electronic background to which the mode is coupled to.

9.1.2. The β''-(BEDT-TTF)$_2$SF$_5$CHFSO$_3$ organic metal

IR vibrational spectroscopy

The IR vibrational measurements with the light polarized along the insulating c-direction in the β''-(BEDT-TTF)$_2$SF$_5$CHFSO$_3$ are presented in Fig. 9.10. As discussed for the isostructutral β''-(BEDT-TTF)$_2$SF$_5$CH$_2$CF$_2$SO$_3$ superconductor, in this frequency range the modes are due to vibrations of the BEDT-TTF molecule or the anion. Again, the modes show typical hardening and sharpen with decreasing temperature. The structure or symmetry does not changes: Neither a significant splitting or changes in position of the anion or molecular modes nor an appearance of additional modes is detected.

The focus here is on the charge sensitive $B_{1u}(\nu_{27})$ vibration which is identified at around 1450 cm^{-1}. The modes at 1408 and 1421 cm^{-1} are assigned to an anion vibrations and a CH$_2$ bending of the BEDT-TTF molecule [66, 174]. The temperature dependence in the region from 1350-1550 cm^{-1} is given in Fig. 9.11. At room temperature a single broad band, identified as $B_{1u}(\nu_{27})$, is observed at 1439 cm^{-1}. On decreasing temperature this band does not split. Mainly the high frequency shoulder of it hardens and sharpens. A fit with two contributions distinguishes the contributions of the $B_{1u}(\nu_{27})$ as high frequency peak and the low frequency part due to other modes like the CH$_2$ bending, etc. These fits are shown (a) for 300 K and (b) for 5 K in Fig. 9.12. The extracted temperature dependence of the parameters is given in Fig. 9.13 (a) for the position and (b) the width. The high frequency peak (green) we assigned to the $B_{1u}(\nu_{27})$ vibration. On cooling to 5 K, the mode hardens by 6 cm^{-1} and sharpens slightly. That is comparable to the hardening in the superconducting sister compound. The low frequency background of the other modes, however, harden less (only 3 cm^{-1}) and in its width is basically temperature independent within the error-bar. That is also the same behavior as in the superconductor.

In terms of spectral weight (Fig. 9.14) the integrated spectral weight of the whole band around 1450 cm^{-1} increases slightly. The $B_{1u}(\nu_{27})$ high frequency peak increases as already seen directly from the growing shoulder in the spectrum. The spectral weight of the low frequency background stays basically constant over the whole temperature range.

In terms of charge redistribution the absence of a splitting means that the charge per molecule stays $0.5e$ in average. This is in perfect agreement with the metallic transport properties. At lowest temperature only, there is the onset of a small shoulder at the high frequency side of the peak (around 1460 cm^{-1}). It is also visible in the fit of the low temperature data in Fig. 9.12 (b): Especially the high frequency peak does not

9. Vibrational spectroscopy

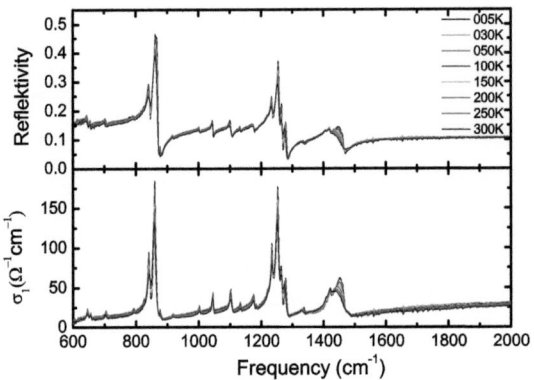

Figure 9.10.: Reflectivity and conductivity of β''-(BEDT-TTF)$_2$SF$_5$CH$_2$CF$_2$SO$_3$ along its insulating c-direction. Temperature evolution and hardening of the BEDT-TTF molecular vibrations and the vibrational modes of the SF$_5$CHFSO$_3$ anion.

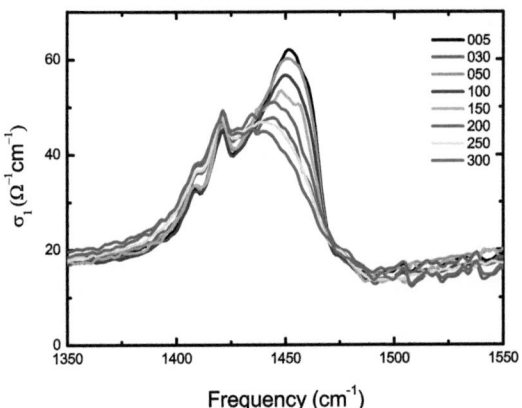

Figure 9.11.: Temperature dependence of the optical conductivity around the charge sensitive $B_{1u}(\nu_{27})$ mode in β''-(BEDT-TTF)$_2$SF$_5$CHFSO$_3$.

9.1. On-site charge distribution in the organic superconductors and metals

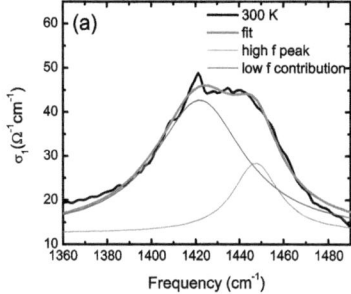
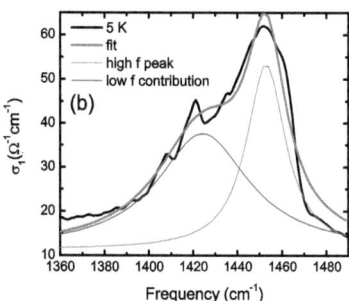

Figure 9.12.: Fit of the contributions to the $B_{1u}(\nu_{27})$ mode in β''-(BEDT-TTF)$_2$-SF$_5$CHFSO$_3$ at (a) 300 K and (b) 5 K.

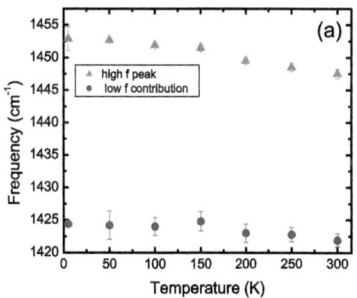
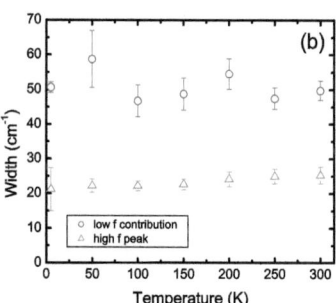

Figure 9.13.: Temperature dependence of the (a) position and (b) width of the contributions to the $B_{1u}(\nu_{27})$ mode in β''-(BEDT-TTF)$_2$SF$_5$CHFSO$_3$.

9. Vibrational spectroscopy

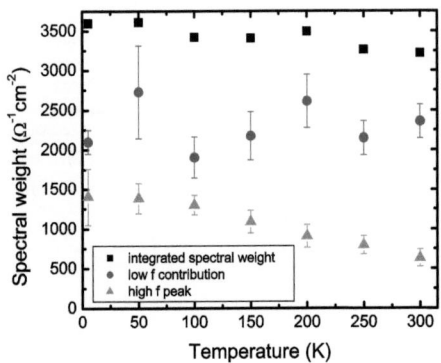

Figure 9.14.: Evolution of spectral weight of the contributions to the $B_{1u}(\nu_{27})$ mode in β''-(BEDT-TTF)$_2$SF$_5$CHFSO$_3$.

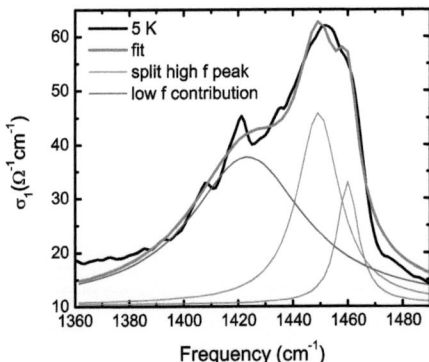

Figure 9.15.: Improved fit of the 5 K spectrum of the $B_{1u}(\nu_{27})$ band in β''-(BEDT-TTF)$_2$-SF$_5$CHFSO$_3$. The high frequency peak is fit here with two contributions.

9.1. On-site charge distribution in the organic superconductors and metals

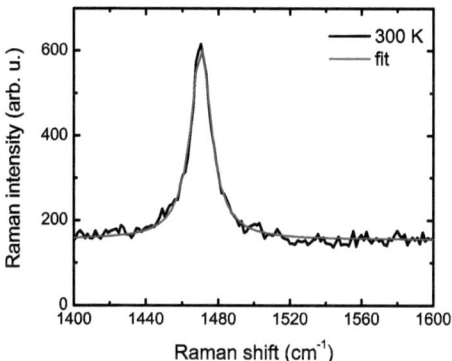

Figure 9.16.: Room temperature Raman spectrum of β''-(BEDT-TTF)$_2$SF$_5$CHFSO$_3$. Only the $A_g(\nu_3)$ mode is present.

describe the high frequency shoulder of the band very well. Fitting the high frequency peak with two contributions(Fig. 9.15) improves the fit quality significantly. It reveals two contributions within the high frequency peak at 1451 and 1460 cm^{-1}. The resulting frequency splitting of $\Delta\nu \approx 9$ cm^{-1} corresponds to a very small charge redistribution of $0.08e$. This could be due to small charge fluctuations within the renormalized metal β''-(BEDT-TTF)$_2$SF$_5$CHFSO$_3$ that become traceable at lowest temperatures. But the are small in comparison to the superconducting sister compound β''-(BEDT-TTF)$_2$-SF$_5$CH$_2$CF$_2$SO$_3$. That shows a large splitting of the $B_{1u}(\nu_{27})$ mode and signs of charge-order are already present at higher temperatures.

Raman spectroscopy

The charge distribution in the β''-(BEDT-TTF)$_2$SF$_5$CHFSO$_3$ is also traced via the symmetric $A_g(\nu_2)$ and $A_g(\nu_3)$ modes using Raman spectroscopy. At room temperature only a single peak at 1470 cm^{-1} is observed as shown in Fig. 9.16. This mode is assigned to the $A_g(\nu_3)$ vibration. In contrast to the superconducting sister compound the peak can be well described using one Lorentzian oscillator. No second side peak is detected in the high frequency shoulder. However, on cooling a second peak appears below 200 K (Fig. 9.17). The fit to the two modes is presented in Fig. 9.18 (a). The second peak at 1493 cm^{-1} is assigned to the $A_g(\nu_2)$ vibration. In that way the mode assignment in the high temperature regime is the same as for the isostructutral superconductor. But here the modes are well separated. Cooling the metallic β''-(BEDT-TTF)$_2$SF$_5$CHFSO$_3$, a

9. Vibrational spectroscopy

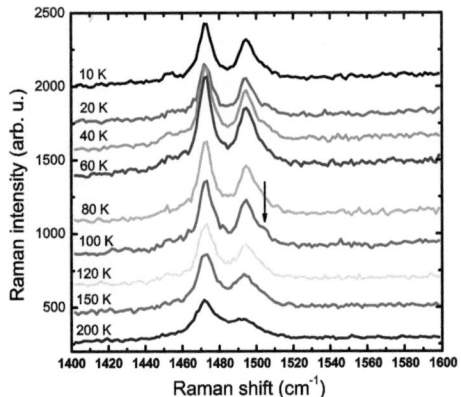

Figure 9.17.: Temperature dependence of the Raman active, charge sensitive $A_g(\nu_2)$ and $A_g(\nu_3)$ modes in β''-(BEDT-TTF)$_2$SF$_5$CHFSO$_3$.

high frequency shoulder in the $A_g(\nu_2)$ at 100 K is visible (small arrow in Fig. 9.17) as some kind of pre-sign. The shoulder is not detected for 80 K but becomes well developed and is fitted for temperatures below 60 K. The example of a fit is shown in Fig. 9.18 (b) for lowest temperatures. This shoulder is attributed to a split contribution of the $A_g(\nu_2)$ mode.

The temperature dependence of (a) the Raman shift and (b) the width all observed peaks of the $A_g(\nu_2)$ and $A_g(\nu_3)$ modes is given in Fig. 9.19. Both modes harden (the $A_g(\nu_3)$ mode from 1470 to 1472 cm^{-1} and the $A_g(\nu_2)$ from 1492 to 1495 cm^{-1}) on cooling to 150 K. For lower temperatures their Raman shifts stay constant. All modes sharpen on cooling. Below 60 K the splitting of the $A_g(\nu_2)$ mode is observed with contributions at 1495 and 1500 cm^{-1}. Both peaks sharpen on cooling and do not show significant changes in their Raman shift. The extracted splitting of $\Delta\nu = 5$ cm^{-1} is small and does not exceed 10 cm^{-1} even within the errorbar. The corresponding charge redistribution is about 0.05e and within the errorbar does not exceed 0.1e.

The observed behavior of no significant splitting is in perfect agreement to the metallic behavior found in transport properties and the results from the IR spectroscopy where the $B_{1u}(\nu_{27})$ charge sensitive band shows no significant splitting. However, at lowest temperatures there is a very small splitting of the $A_g(\nu_2)$ in the Raman spectrum and in IR spectroscopy a very small splitting of the $B_{1u}(\nu_{27})$ mode is detected. The cor-

9.1. On-site charge distribution in the organic superconductors and metals

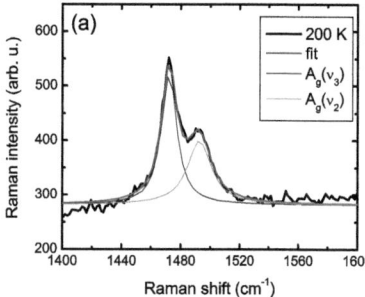 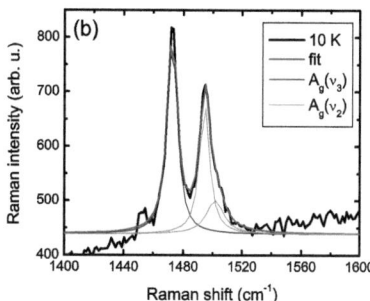

Figure 9.18.: Numerical fit to the $A_g(\nu_2)$ and $A_g(\nu_3)$ modes in β''-(BEDT-TTF)$_2$-SF$_5$CHFSO$_3$ at (a) 200 K and (b) 10 K. In the low temperature regime a split $A_g(\nu_2)$ is found.

responding charge redistributions are in the same range (0.05-0.1e) for both methods. Thus in both cases, IR and Raman, there is a trace for a slight disturbance of the homogeneous charge distribution. The splittings corresponds to a small charge redistribution and the small spectral weight of the features shows that the main part of the system has equally distributed charges. Only a small fraction might exhibit fluctuating charge order patterns. That is in agreement with the picture of a slightly renormalized metal. At the same temperature (around 100 K) when the first traces of the slight splitting are observed, also a tiny Raman mode starts to develop at the low frequency shoulder of the $A_{1g}(\nu_3)$ mode at around 1450 cm^{-1}. The origin of this mode is unclear so far. In the IR spectroscopy no corresponding traces can be seen; there it is right the region of the $B_{1u}(\nu_{27})$ mode.

9.1.3. The β-(EDT-TTF)$_4$[Hg$_3$I$_8$]$_{(1-x)}$ organic superconductor

To further investigate the charge order influence on the superconductors, the vibrational properties of the organic β-(EDT-TTF)$_4$[Hg$_3$I$_8$]$_{(1-x)}$ superconductor with $T_c = 8.1$ K are measured. Since the EDT-TTF molecule is asymmetric most of its vibrational modes have a dipole moment and are IR-active. The spectrum along the insulating direction is given in Fig. 9.20. Comparing the resolved modes with results from BEDT-TTF based salts and theoretical calculations from the similar salt (EDT-DTDSF)$_4$Hg$_3$I$_8$, where two sulfur atoms are replaced with selenium, all appearing modes can be assigned (for details see diploma thesis of C. Clauss [86]). No distinct changes of the anion modes are detected on cooling: The structure does not change. The charge sensitive C=C double bond vibration, equivalent to the $B_{1u}(\nu_{27})$ mode in BEDT-TTF salts, is observed at

9. Vibrational spectroscopy

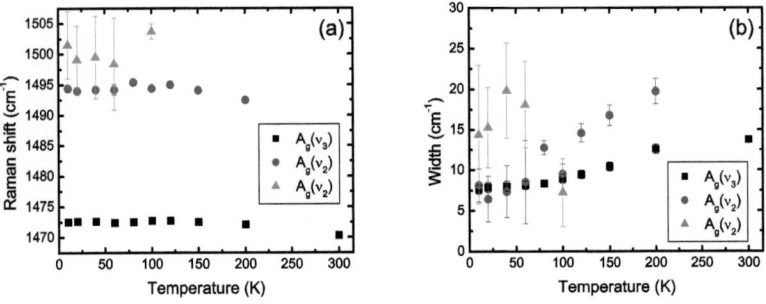

Figure 9.19.: Temperature dependence of the (a) position and (b) width of the contributions to the $A_g(\nu_2)$ and $A_g(\nu_3)$ modes in β''-(BEDT-TTF)$_2$SF$_5$CHFSO$_3$.

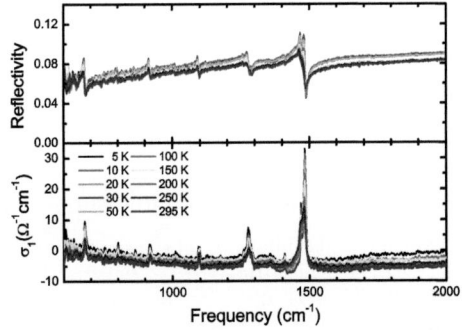

Figure 9.20.: Reflectivity and optical conductivity measured with the light polarized along the insulating direction in β-(EDT-TTF)$_4$[Hg$_3$I$_8$]$_{(1-x)}$.

9.1. On-site charge distribution in the organic superconductors and metals

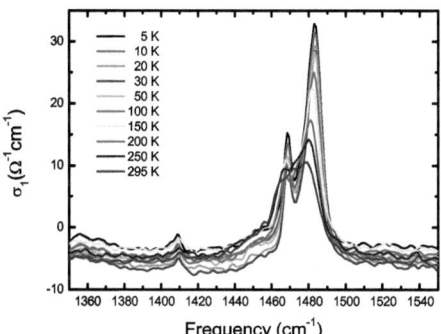

Figure 9.21.: Temperature dependence of the optical conductivity around the charge sensitive central C=C stretching vibration in β-(EDT-TTF)$_4$[Hg$_3$I$_8$]$_{(1-x)}$.

around 1475 cm^{-1} for β-(EDT-TTF)$_4$[Hg$_3$I$_8$]$_{(1-x)}$. Its temperature dependence is plotted in Fig. 9.21. At room temperature there is a broad single band which harden on cooling and split for temperatures below 250 K. On cooling mainly the high frequency peak enhances. The shape of the band at room temperature already suggests that there are two contributions. A fit of the (a) room temperature spectrum and (b) at 5 K are given in Fig. 9.22. These two peaks are assumed to be the contributions to the charge sensitive C=C double bond vibration. The fit describes the spectra well. A very small side-tail at low frequencies between 1440-1460 cm^{-1} could be some small background and is neglected here. The temperature dependence of the parameters is given in Fig. 9.23 (a) for the position and (b) the width of the two peaks. On cooling the high-frequency peak hardens linearly from 1480 to 1483 cm^{-1}. Below 100 K the slope decreases and below 30 K the mode even softens. On the other hand the low-frequency peak shows a softening from 1470 cm^{-1} at 250 K down to 1468 cm^{-1} at 30 K, then the mode hardens. That means the splitting between the modes increases on cooling to 50-30 K, then it decreases for lower temperatures. The width of the high frequency peaks narrows slightly and linear on cooling while the low frequency peak significantly narrows linear down to about 20 K; below it slightly widens, again.

The spectral weight of the peaks, given in Fig. 9.24, shows that the weight of the whole band is conserved within the error-bars. The spectral weight distribution between the peaks reveals that on cooling to 30 K the low frequency peak looses spectral weight to the high frequency peak. Cooling further shifts back a small fraction of spectral weight.

To understand the detailed behavior, as described above, more knowledge about the

9. Vibrational spectroscopy

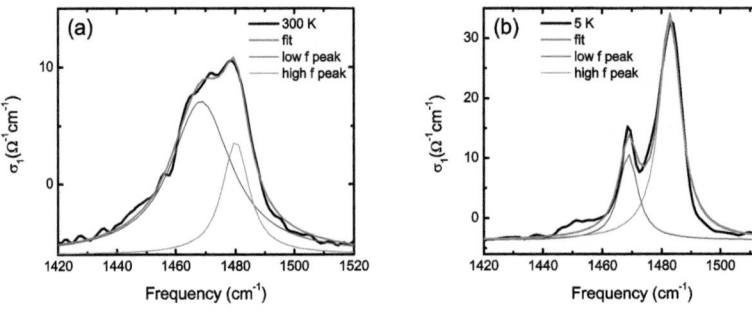

Figure 9.22.: Numerical fit of the contributions to the charge sensitive central C=C stretching vibration in β-(EDT-TTF)$_4$[Hg$_3$I$_8$]$_{(1-x)}$ at (a) 300 K and (b) 5 K.

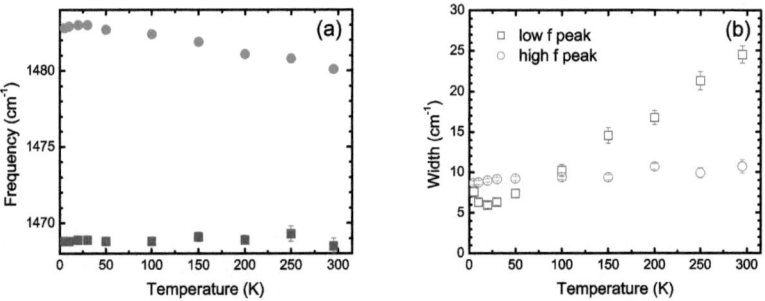

Figure 9.23.: Temperature dependence of the (a) position and (b) width of the contributions to the charge sensitive central C=C stretching vibration in β-(EDT-TTF)$_4$-[Hg$_3$I$_8$]$_{(1-x)}$.

9.1. On-site charge distribution in the organic superconductors and metals

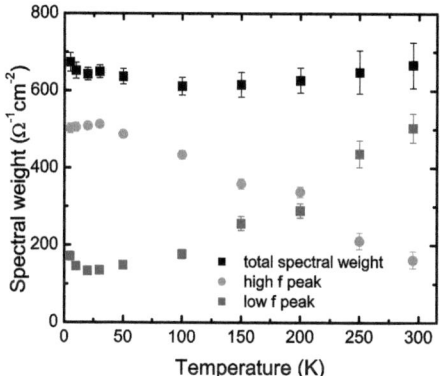

Figure 9.24.: Evolution of spectral weight of the contributions to the charge sensitive central C=C stretching vibration in β-(EDT-TTF)$_4$[Hg$_3$I$_8$]$_{(1-x)}$.

actual charge distribution within the EDT-TTF is needed. However, it is similar to the behavior measured for the β''-(BEDT-TTF)$_2$SF$_5$CH$_2$CF$_2$SO$_3$ superconductor. A broad band at high temperature significantly splits on cooling. With decreasing temperature, for the β-(EDT-TTF)$_4$[Hg$_3$I$_8$]$_{(1-x)}$ already at room temperature, a splitting of the mode sets in. That represents the onset of a charge redistribution between the molecular sites. Assuming the same correspondence of 105 cm^{-1}/e as for the BEDT-TTF molecule the observed splitting of $\Delta\nu = 14.6$ cm^{-1} evidences a charge redistribution of $0.14e$ between the sites.

The agreement in the behavior of the two superconductors, the β''-(BEDT-TTF)$_2$-SF$_5$CH$_2$CF$_2$SO$_3$ and β-(EDT-TTF)$_4$[Hg$_3$I$_8$]$_{(1-x)}$, is remarkable. Both are basically metallic as characterized by their transport properties and both show a small charge redistribution between the sites of about $0.2e$ as detected by vibrational spectroscopy. Also the spectral weight of the split contributions is of equal amount as the broad mode at room temperature. This shows an increasing amount of temporary charge-ordered sites in expense of the equally distributed sites on cooling. That suggests an increased influence of charge fluctuations in the charge-ordered metallic state.

9. Vibrational spectroscopy

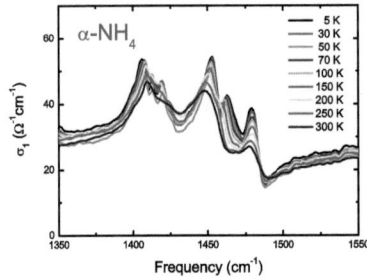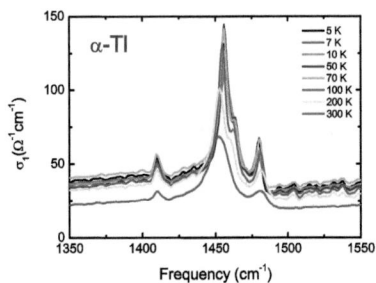

Figure 9.25.: Temperature dependence of the optical conductivity around the charge sensitive $B_{1u}(\nu_{27})$ mode at about 1450 cm^{-1} in α-(BEDT-TTF)$_2$-NH$_4$Hg(SCN)$_4$ and α-(BEDT-TTF)$_2$TlHg(SCN)$_4$. The modes at 1410 and 1480 cm^{-1} are assigned to CH$_2$ modes.

9.1.4. The α-(BEDT-TTF)$_2$NH$_4$Hg(SCN)$_4$ and α-(BEDT-TTF)$_2$TlHg(SCN)$_4$ organic metals

As additional superconductor-metal pair two members of the 1/4-filled α-(BEDT-TTF)$_2$$M$Hg(S family are investigated. The superconducting α-(BEDT-TTF)$_2$NH$_4$Hg(SCN)$_4$ ($T_c = 1$ K) and its metallic sister compound α-(BEDT-TTF)$_2$TlHg(SCN)$_4$ are known to be close to a charge-order transition.

IR vibrational spectroscopy

The $B_{1u}(\nu_{27})$ mode, investigated with the light along the insulating direction, is given in Fig. 9.25 for (a) the superconductor α-(BEDT-TTF)$_2$NH$_4$Hg(SCN)$_4$ and (b) the metallic α-(BEDT-TTF)$_2$TlHg(SCN)$_4$. At room temperature the $B_{1u}(\nu_{27})$ mode is located as single band at 1448 cm^{-1} for α-(BEDT-TTF)$_2$NH$_4$Hg(SCN)$_4$ and 1452 cm^{-1} for α-(BEDT-TTF)$_2$TlHg(SCN)$_4$. The modes around 1408 and 1480 cm^{-1} are CH$_2$ modes. On cooling the $B_{1u}(\nu_{27})$ band starts to sharpen and harden in both cases. While for the superconductor α-(BEDT-TTF)$_2$NH$_4$Hg(SCN)$_4$ a distinct splitting of the mode below 200 K starts to develop, a shoulder appears for the metallic α-(BEDT-TTF)$_2$TlHg(SCN)$_4$ compound. In both cases these side peaks carry only about a tenth of spectral weight of the main peak. The integrated spectral weight (Fig. 9.26) contributing to the $B_{1u}(\nu_{27})$ mode increases in both systems. (In the Tl-compound the data point at 300 K most likely is too low. The optical conductivity at that temperature seems to be unusual low compared to the other temperatures. However position and width of the peaks are not affected by that.)

9.1. On-site charge distribution in the organic superconductors and metals

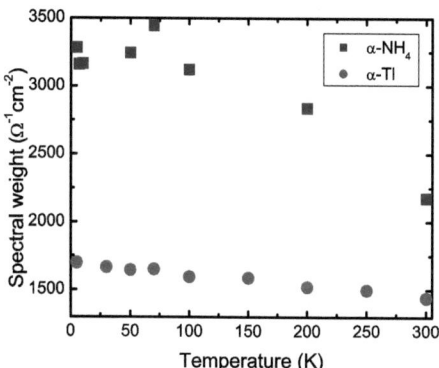

Figure 9.26.: Integrated spectral weight of the two contributions of the $B_{1u}(\nu_{27})$ vibration in α-(BEDT-TTF)$_2$NH$_4$Hg(SCN)$_4$(red) and α-(BEDT-TTF)$_2$-TlHg(SCN)$_4$(blue) as function of temperature.

The detailed behavior of the individual peaks contributing to the $B_{1u}(\nu_{27})$ mode in both compounds is followed by fitting them with Lorentzian oscillators. Their parameters are given in Fig. 9.27 for (a) the position and (b) the width of the peaks in α-(BEDT-TTF)$_2$-TlHg(SCN)$_4$ (blue) and α-(BEDT-TTF)$_2$NH$_4$Hg(SCN)$_4$ (red). Clearly the hardening and splitting of the mode is seen. The splitting between the modes (Fig. 9.26) enhances from 10 up to 13 cm^{-1} for α-(BEDT-TTF)$_2$NH$_4$Hg(SCN)$_4$ on cooling. On the other hand in α-(BEDT-TTF)$_2$TlHg(SCN)$_4$ the splitting stays basically constant at 9 cm^{-1} till 100 K. Then on cooling below it reduces to about 7 cm^{-1}. At lowest temperatures, below 10 K, a small increase of the splitting seems to set in.

The splitting of $\Delta\nu = 9 - 10$ cm^{-1} right at the onset of the charge redistribution corresponds to a charge difference of about 0.1e at 200 K between the sites. For the superconducting α-(BEDT-TTF)$_2$NH$_4$Hg(SCN)$_4$ it increases on cooling to 0.13e with a tendency to increase more on further cooling. On the other hand in the metallic α-(BEDT-TTF)$_2$TlHg(SCN)$_4$ it reduces to a charge disproportionation of 0.07e at about 5 K.

Comparing the two systems, the superconductor α-(BEDT-TTF)$_2$NH$_4$Hg(SCN)$_4$ shows the larger splitting and therefore larger charge redistribution between the sites. This disproportionation even increases to lowest temperatures. Compared to the other superconductors, the charge redistribution here is low. That could explain the low T$_c \approx 1$ K. In the α-(BEDT-TTF)$_2$TlHg(SCN)$_4$ the charge redistribution is smaller and decreases on cooling. That puts the system more to the metallic in the phase diagram. At low-

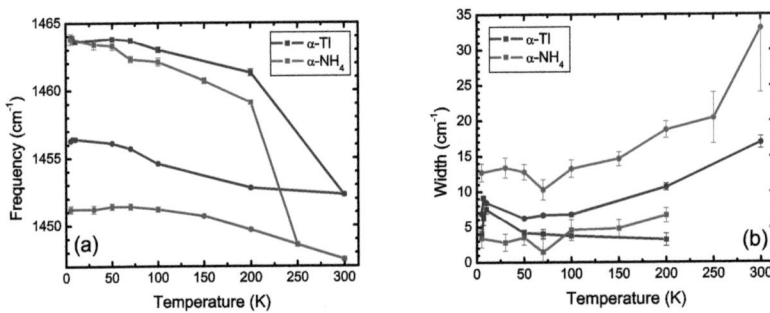

Figure 9.27.: Temperature dependence of (a) postion and (b) width of the contributions to the $B_{1u}(\nu_{27})$ mode in α-(BEDT-TTF)$_2$NH$_4$Hg(SCN)$_4$ (red) and α-(BEDT-TTF)$_2$TlHg(SCN)$_4$ (blue). Here only the peaks around 1450 cm^{-1} are taken into account and assumed to contribute.

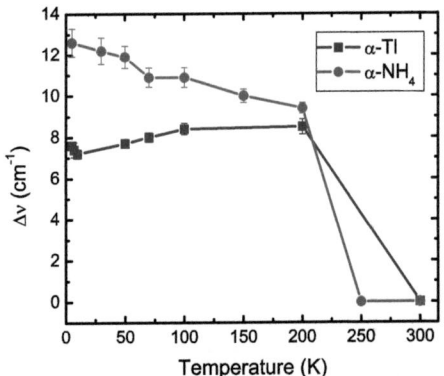

Figure 9.28.: Splitting between the two contributions of the $B_{1u}(\nu_{27})$ vibration in α-(BEDT-TTF)$_2$NH$_4$Hg(SCN)$_4$ (red) and α-(BEDT-TTF)$_2$TlHg(SCN)$_4$ (blue) as function of temperature.

9.1. On-site charge distribution in the organic superconductors and metals

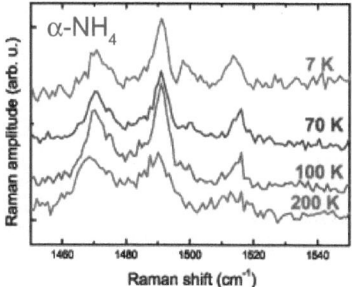
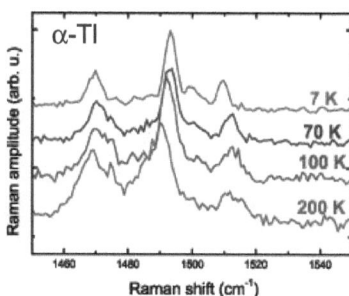

Figure 9.29.: Temperature dependence of the Raman active, charge sensitive ν_2 mode in α-(BEDT-TTF)$_2$NH$_4$Hg(SCN)$_4$(a) and α-(BEDT-TTF)$_2$TlHg(SCN)$_4$(b). Measurements performed by N. Drichko and M. Masino at Dep. Chimica G.I.A.F. and INSTM-UdR Parma University.

est temperatures, below 50 K, a reentrant like behavior to a more metallic state is observed. Further, in contrast to the β''-(BEDT-TTF)$_2$SF$_5$CH$_2$CF$_2$SO$_3$ and β-(EDT-TTF)$_4$[Hg$_3$I$_8$]$_{(1-x)}$ superconductors the mode splitting and therefore the charge redistribution does not stay constant for the α-systems.

Raman spectroscopy

To gain additional information about the charge redistribution Raman spectra of the superconductor α-(BEDT-TTF)$_2$NH$_4$Hg(SCN)$_4$ and the metal α-(BEDT-TTF)$_2$TlHg(SCN)$_4$ are shown in Fig. 9.29 (a) and (b) respectively. In both compounds the $A_g(\nu_3)$ mode is located around 1470 cm^{-1} while the $A_g(\nu_2)$ contributions are located around 1495 cm^{-1} and 1510 cm^{-1}. The respective center frequencies are given in Fig. 9.30 for the ν_3 (open circles) and ν_2 contributions (filled symbols). The superconductor α-(BEDT-TTF)$_2$NH$_4$Hg(SCN)$_4$ is shown in red while the α-(BEDT-TTF)$_2$TlHg(SCN)$_4$ metal is given in blue. The $A_g(\nu_2)$ clearly splits in both compounds. For the superconductor the corresponding charge redistribution accounts for 0.25e while in the metal its significantly lower ($\Delta\rho = 0.18e$) at lowest temperatures. While for the superconductor α-(BEDT-TTF)$_2$NH$_4$Hg(SCN)$_4$ the splitting starts to develop on decreasing temperature and stays constant at lowest temperatures, the metallic α-(BEDT-TTF)$_2$TlHg(SCN)$_4$ shows a decrease of splitting to lowest temperatures. That corresponds nicely to the results of the the IR active ν_{27} mode. An increased splitting to lower temperatures was found in the α-(BEDT-TTF)$_2$NH$_4$Hg(SCN)$_4$. The reentrant behavior in the α-(BEDT-TTF)$_2$TlHg(SCN)$_4$ is confirmed. The extracted charge redistribution values in Raman

9. Vibrational spectroscopy

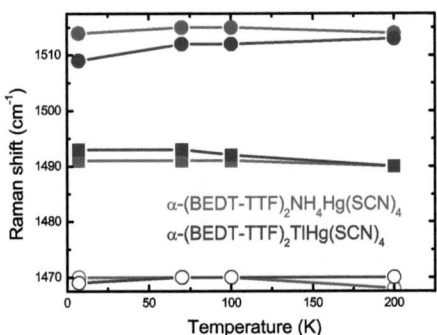

Figure 9.30.: Temperature dependence of the Raman shift of the charge sensitive, Raman active modes in α-(BEDT-TTF)$_2$NH$_4$Hg(SCN)$_4$ (red) and α-(BEDT-TTF)$_2$-TlHg(SCN)$_4$ (blue). From N. Drichko and M. Masino.

are slightly higher compared to the ones from IR. But the superconductor has the higher redistribution in both cases.

These results are also in line with the observations from the β''-systems. The superconductors show a large and significant charge disproportionation while in the metals no or only weak traces of charge redistribution are present. However, Ref. [78] proposes α-(BEDT-TTF)$_2$TlHg(SCN)$_4$ to be the stronger correlated system compared with the superconducting α-(BEDT-TTF)$_2$NH$_4$Hg(SCN)$_4$ (c.f. Fig. 2.23). On the other hand the results of the vibrational analysis show the α-(BEDT-TTF)$_2$TlHg(SCN)$_4$ on the more metallic side. In the optical measurements [78] also the reentrant behavior to the more metallic side is observed in the direction parallel to the stacks. The placement of the α-(BEDT-TTF)$_2$TlHg(SCN)$_4$ on the more correlated side in Ref. [78] was considered due to the strong absolute increase of the correlation features. Based on the relative strength of metallic and correlation properties and on the results here, the α-(BEDT-TTF)$_2$TlHg(SCN)$_4$ would be placed to the metallic side of the phase diagram compared to the higher correlated superconductor.

9.2. The metal insulator transition in α-(BEDT-TTF)$_2$I$_3$ and θ-(BEDT-TTF)$_2$RbZn(SCN)$_4$

The organic conductors θ-(BEDT-TTF)$_2$RbZn(SCN)$_4$ and α-(BEDT-TTF)$_2$I$_3$ both undergo a very well characterized metal-insulator transition on cooling. Therefore a signif-

9.2. The metal insulator transition in α-(BEDT-TTF)$_2$I$_3$ and θ-(BEDT-TTF)$_2$RbZn(SCN)$_4$

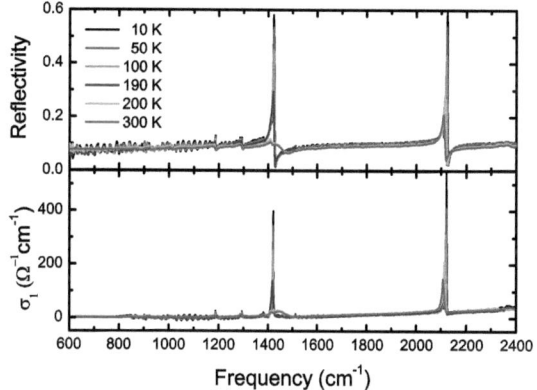

Figure 9.31.: Reflectivity and optical conductivity of θ-RbZn along the insulating b-axis.

icant splitting in the $B_{1u}(\nu_{27})$ is expected and makes the systems to ideal candidates to prove the concept of the on-site charge determination via IR vibrational spectroscopy. The charge-order transition in the θ-(BEDT-TTF)$_2$RbZn(SCN)$_4$ is accompanied by a structural transition where the unit cell is doubled. That makes the α-(BEDT-TTF)$_2$I$_3$ system in particular interesting to compare since below the metal insulator transition it has the same unit cell as α-(BEDT-TTF)$_2$I$_3$. For the latter no structural transition is found at the phase transition. The comparison of both gives insight into how the metal insulator takes place, to what extent the lattice is involved, and how this manifests in the vibrational properties.

9.2.1. The charge order transition in θ-(BEDT-TTF)$_2$RbZn(SCN)$_4$

For the θ-(BEDT-TTF)$_2$RbZn(SCN)$_4$ compound it is known that on slow cooling the charge order transition is accompanied by a structural phase transition which both take place at 190 K [63, 77]. Here IR vibrational spectroscopy probes the metal insulator transition by investigating the behavior of the $B_{1u}(\nu_{27})$ vibrational mode. The structural transition should be evident in anion as well as in vibrational modes of the BEDT-TTF molecule. The whole spectral range investigated perpendicular to the conducting plane is presented in Fig. 9.31. The modes below 1300 cm^{-1} could be easily assigned using literature data as given in Tab. 9.1. However, the most significant temperature dependence is seen in two distinct features. The sharp mode at $\sim$ 1420 cm^{-1} and the

9. Vibrational spectroscopy

Frequency [cm^{-1}]		ν_i	Symmetry	Type
exp.	lit.			
1294	1284	ν_{29}	B_{1u}	CH$_2$ wagging
1183	1175	ν_{21}	B_{1g}	CH$_2$ twisting
1009	997	ν_{47}	B_{2u}	CH$_2$ wagging
923	918	ν_{31}	B_{1u}	CH$_2$ rocking

Table 9.1.: Experimentally obtained mode frequencies, their literature values, assignment and symmetry (literature values taken from [93, 97, 175]).

one at $\sim$ 2120 cm^{-1}. The first one is part of the $B_{1u}(\nu_{27})$ vibrational mode and will be discussed later. The second is assigned to the C=N double bond stretching mode of SCN$^-$ in the anion [32]. The mode hardens with decreasing temperature and below the transition at 190 K an additional mode can be distinguished at 2107 cm^{-1} (Fig. 9.32). That is a direct sign of the structural transition. The second peak appears due to symmetry break within the unit cell. A second effect of the structural transition is seen in the $B_{1g}(\nu_{21})$ and $B_{1u}(\nu_{29})$ modes of the BEDT-TTF molecule. Below 190 K neighboring peaks appear as shown in Fig. 9.33, also due to the symmetry break at the structural phase transition.

The temperature dependence of the $B_{1u}(\nu_{27})$ mode in θ-(BEDT-TTF)$_2$RbZn(SCN)$_4$ is presented in Fig. 9.34. At high temperatures the mode is a broad feature around 1447 cm^{-1} corresponding to an equal charge distribution of 0.5e per site. The peak at 1413 cm^{-1} is assigned to the CH$_2$ bending mode giving rise to a low frequency background similar as for the β''-(BEDT-TTF)$_2$SF$_5$CH$_2$CF$_2$SO$_3$ superconductor. At the metal insulator transition at 190 K the broad $B_{1u}(\nu_{27})$ mode vanishes and three peaks appear at 1420 (a, very strong), 1458 (b, weak) and 1513 cm^{-1} (c, weak). The interpretation for the mode appearing at 1382 cm^{-1} is not understood so far. But it has to be of a different origin than the charge order or structural transition, since it only appears for temperatures below 100 K.

Judged on its position, the mode marked with b could be remnant of the broad band above the transition. The frequency shift is about 10 cm^{-1} and could be due to the structural phase transition. That means even below the metal insulator transition molecular sites with an average charge of 0.5e are present. The modes a and c represent the charge rich and charge poor sites, respectively. A detailed description of their position and width extracted from a fit is given in Fig. 9.35. At the phase transition at 190 K the frequency of the broad center mode (black squares) jumps and the low (red circles) and high (green triangular) frequency peaks of the split mode appear. Below the phase transition all three peaks slightly harden on cooling. For the center mode the width reduces from 45 to 11 cm^{-1} at the phase transition. The spectral weight (Fig. 9.36) shows a large transfer of spectral weight to the low frequency peak, while the center peak and

9.2. The metal insulator transition in α-(BEDT-TTF)$_2$I$_3$ and θ-(BEDT-TTF)$_2$RbZn(SCN)$_4$

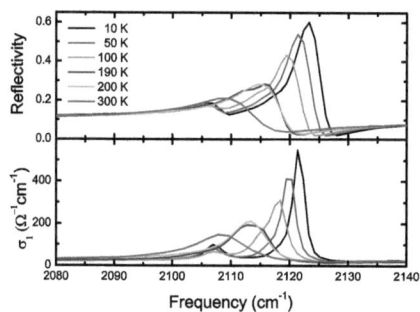

Figure 9.32.: Splitting of the C=N stretching mode of the RbZn(SCN)$_4$ anion at 190 K indicating the stuctural phase transition.

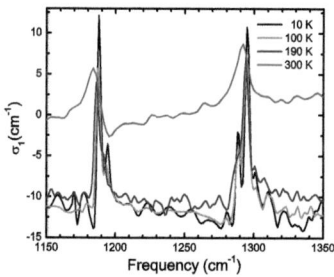

Figure 9.33.: Modes ν_{21} and ν_{29} above and below the transition temperature (190 K) of θ-(BEDT-TTF)$_2$RbZn(SCN)$_4$. The appearance of the side peaks shows the structural phase transition. Number of temperatures reduced for clarity.

9. Vibrational spectroscopy

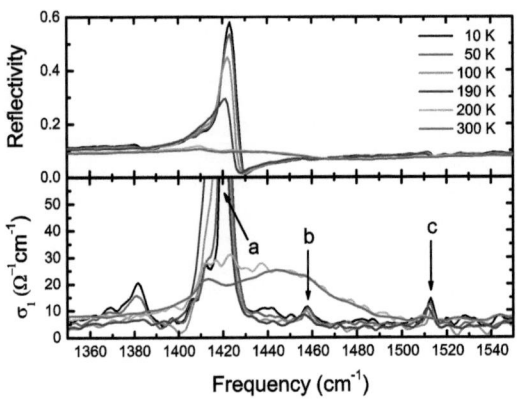

Figure 9.34.: Temperature dependent development of the $B_{1u}(\nu_{27})$ mode in θ-(BEDT-TTF)$_2$RbZn(SCN)$_4$.

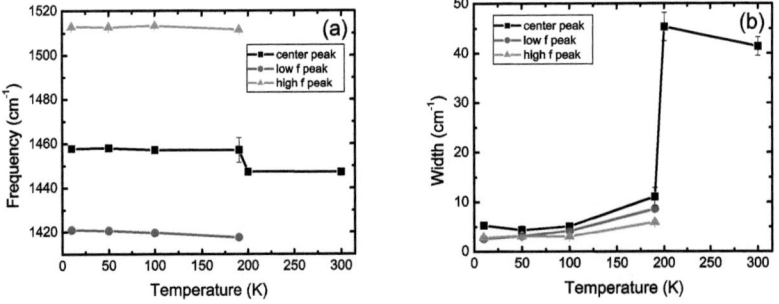

Figure 9.35.: Temperature dependence of (a) postion and (b) width of the contributions to the $B_{1u}(\nu_{27})$ mode in θ-(BEDT-TTF)$_2$RbZn(SCN)$_4$.

9.2. The metal insulator transition in α-(BEDT-TTF)$_2$I$_3$ and θ-(BEDT-TTF)$_2$RbZn(SCN)$_4$

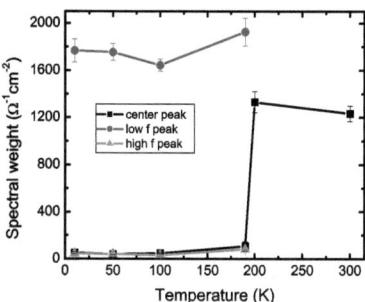

Figure 9.36.: Spectral weight of the three contributions of the $B_{1u}(\nu_{27})$ mode of θ-(BEDT-TTF)$_2$RbZn(SCN)$_4$ as function of temperature.

high frequency peak become very small. At the transition the total spectral weight is obviously not conserved.

The frequency splitting between the high and low frequency peak of $\Delta\nu = 92$ cm^{-1} corresponds to a charge redistribution of $0.88e$ assuming a slope of 105 cm$^{-1}e^{-1}$ as for the systems investigated before. This value is above the value of charge disproportionation obtained with different experimental techniques. NMR, X-ray studies, or Raman measure about $0.6e$ [32, 63, 65, 74, 77, 150, 176]. Using the steep slope of 140 cm$^{-1}e^{-1}$ in the charge-frequency dependence, as e.g. used for θ-phases in Ref. [100], $\Delta\rho$ is calculated to $0.66e$, which is in agreement with results obtained from other techniques.

Unclear so far are the large differences in the spectral weight of the $B_{1u}(\nu_{27})$ representing the contribution from the differently charged molecules as seen in the a and c peaks in Figs. 9.34 and 9.36. It could be due to the large charge redistribution but also due to the structural transition. Calculations on that issue were performed by A. Girlando and are presented later in Fig. 9.47. They indicate that charge order leads to strong enhancement of the low frequency peak. That suggests that the charge order changes the dipole moment of the mode significantly and the light can couple much stronger to it.

9.2.2. The metal insulator transition in the $h8$-α-(BEDT-TTF)$_2$I$_3$ and $d8$-α-(BEDT-TTF)$_2$I$_3$ system

The α-(BEDT-TTF)$_2$I$_3$ system is also heavily investigated, especially with focus on the metal-insulator transition at 135 K. X-Ray investigations do not show any structural changes taking place at that transition but there are hints for structural changes at lower temperatures [149, 177]. Here we probe the charge redistribution within the system

9. Vibrational spectroscopy

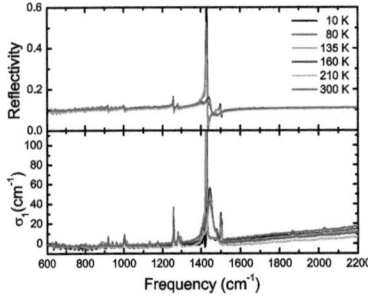
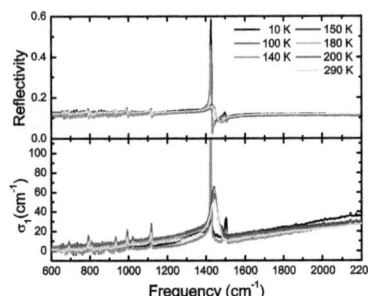

Figure 9.37.: Reflectivity and conductivity spectra of $h8$-α-(BEDT-TTF)$_2$I$_3$ (left) and the deuterated $d8$-α-(BEDT-TTF)$_2$I$_3$ system (right).

during cooling to follow the metal-insulator transition and also to search for fingerprints of structural changes at very low temperatures. Since the unit cells of the α-(BEDT-TTF)$_2$I$_3$ is similar to the doubled unit cell of θ-(BEDT-TTF)$_2$RbZn(SCN)$_4$ below its metal-insulator transition its worthwhile to compare the transition in both systems to gain further insight into the open questions like the lattice or charge density influence on the spectra. For better understanding and assignment of the spectra also the deuterated $d8$-α-(BEDT-TTF)$_2$I$_3$ system is measured.

The reflectivity and conductivity spectra along the insulating direction of the $h8$-α-(BEDT-TTF)$_2$I$_3$ and the deuterated $d8$-α-(BEDT-TTF)$_2$I$_3$ system are given in Fig. 9.37. They are similar to the θ-(BEDT-TTF)$_2$RbZn(SCN)$_4$ response. The main feature is the significant change of the $B_{1u}(\nu_{27})$ at the phase transition. To make a proper assignment of the modes, the results of the $h8$ system are compared with the deuterated $d8$ system. For the room temperature spectra the comparison is presented in Fig. 9.38. All modes which involve hydrogen atoms shift to lower frequencies for the deuterated system. So a final assignment of the modes is done as given in Tab. 9.2.

The temperature dependent behavior of the vibrational spectra of the BEDT-TTF molecule is basically the same for the hydrogenated and the deuterated compound. The enlarged view of the region around the $B_{1u}(\nu_{27})$ is shown for $h8$-α-(BEDT-TTF)$_2$I$_3$ in Fig. 9.39, for instance. The charge sensitive $B_{1u}(\nu_{27})$ mode for the two compounds is compared at room temperature and lowest temperature in Fig. 9.40. A split mode is present already at room temperature. A broad mode at 1443 cm^{-1} but also a small hump around 1476 cm^{-1} is assigned to the $B_{1u}(\nu_{27})$ vibration corresponding to charge redistribution of $0.23e$ between the site. Transport measurements reveal a metallic state above 135 K. Therefore a charge-ordered metallic state close to a charge order transition is concluded. Alternatively the charge order pattern could be spread over the three

9.2. The metal insulator transition in α-(BEDT-TTF)$_2$I$_3$ and θ-(BEDT-TTF)$_2$RbZn(SCN)$_4$

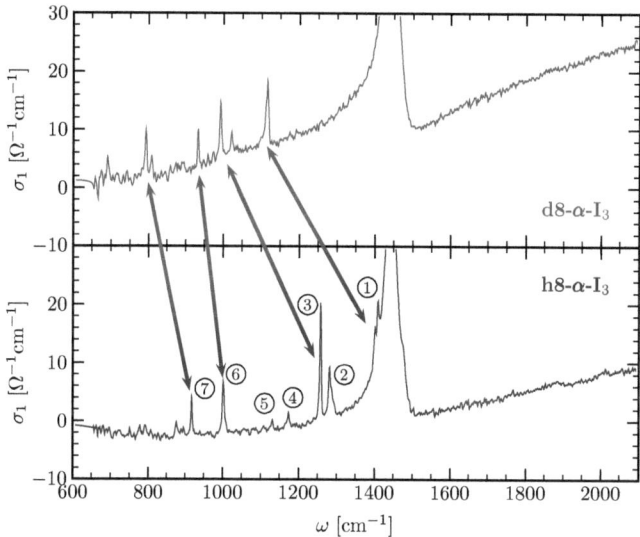

Figure 9.38.: Comparison of the room temperature spectra of $d8$- and $h8$-α-(BEDT-TTF)$_2$-I$_3$. From [86].

marker	h8-α-I$_3$		d8-α-I$_3$		ν_i	Type
	exp.	lit.	exp.	lit.		
(1)	1410	1420	1110	1118	$B_{1u}(\nu_{28})$	CH$_2$-, CD$_2$ bending
(2)	1284	1282	1021	1011	$B_{1u}(\nu_{29})$	CH$_2$-, CD$_2$ bending
(3)	1258	1261	993	1013	$B_{2u}(\nu_{46})$	CH$_2$-, CD$_2$ bending
(4)	1174	1175	—	—	$B_{1g}(\nu_{21})$	CH$_2$-, CD$_2$ twisting
(5)	1132	1125	—	—	$B_{3u}(\nu_{67})$	CH$_2$-, CD$_2$ twisting
(6)	1001	997	933	931	$B_{2u}(\nu_{47})$	CH$_2$-, CD$_2$ wagging
(7)	917	918	792	793	$B_{1u}(\nu_{31})$	SCH$_2$-, SCD$_2$ stretching + CH$_2$-, CD$_2$ wagging

Table 9.2.: Experimental and literature frequencies and their symmetry and type of the modes seen in $h8$- and $d8$-α-(BEDT-TTF)$_2$I$_3$. From [86].

9. Vibrational spectroscopy

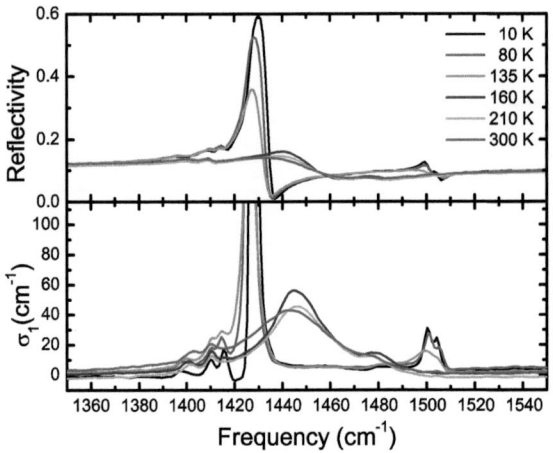

Figure 9.39.: Behavior of the $B_{1u}(\nu_{27})$ mode in $h8$-α-(BEDT-TTF)$_2$I$_3$ at temperatures above and below the transition temperature.

different lattice sites present in the α-systems. The deuterated $d8$-α-(BEDT-TTF)$_2$I$_3$ does not show the second peak around 1480 cm^{-1} at room temperature. On cooling a second contribution becomes visible below 200 K. At lowest temperatures, well below the phase transition, the split $B_{1u}(\nu_{27})$ mode is present with four contributions. They appear in pairs (Fig. 9.40): In the $h8$-system one pair at (a) 1415 and (b) 1428 cm^{-1} ((a') 1416 and (b') 1425 cm^{-1} for the $d8$-system) and the second pair at (c) 1500 and (d) 1504 cm^{-1} (same positions (c',d') in the $d8$-system). These would correspond to a charge redistribution of about $0.7e$ from the (a,d) splitting and $0.55e$ from the (b,c) splitting. All other vibrational modes are of different origin and are significantly shifted in the $d8$-compound (Fig. 9.38).

The detailed change in the spectrum (Fig. 9.41) over the transition in $h8$-α-(BEDT-TTF)$_2$I$_3$ is investigated for temperatures around the transition at 135 K. The splitting of the $B_{1u}(\nu_{27})$ mode is similar to the one observed in the θ-(BEDT-TTF)$_2$RbZn(SCN)$_4$: A broad band above the transition splits into well separated bands of different spectral weight below the transition. The difference here is that two contributions are present above and four contributions are present below the transition. This was verified by assigning the modes in comparison to the $d8$-α-(BEDT-TTF)$_2$I$_3$. As concluded from

9.2. The metal insulator transition in α-(BEDT-TTF)$_2$I$_3$ and θ-(BEDT-TTF)$_2$RbZn(SCN)$_4$

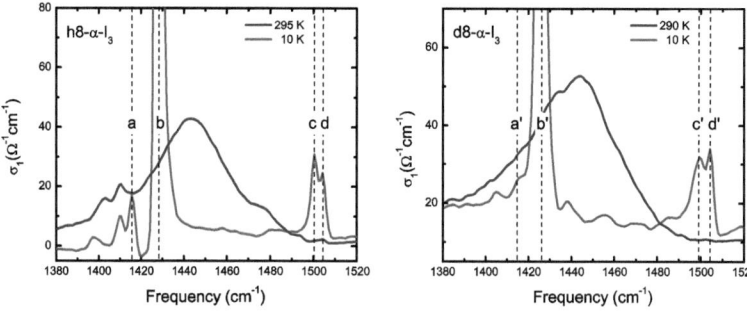

Figure 9.40.: Comparison of the conductivity spectra for the $B_{1u}(\nu_{27})$ feature of $h8$-(left) and $d8$-α-(BEDT-TTF)$_2$I$_3$ (right) for room temperature and very low temperature.

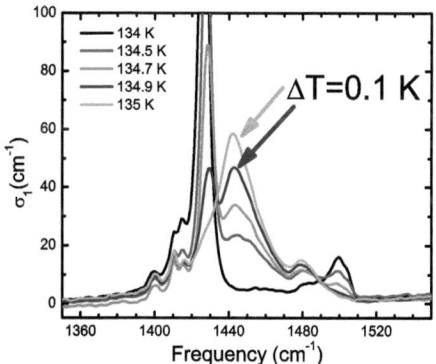

Figure 9.41.: The detailed change of the $B_{1u}(\nu_{27})$ mode on the transition in $h8$-α-(BEDT-TTF)$_2$I$_3$. The transition from the broad band to the sharp features takes place within 1 K.

149

9. Vibrational spectroscopy

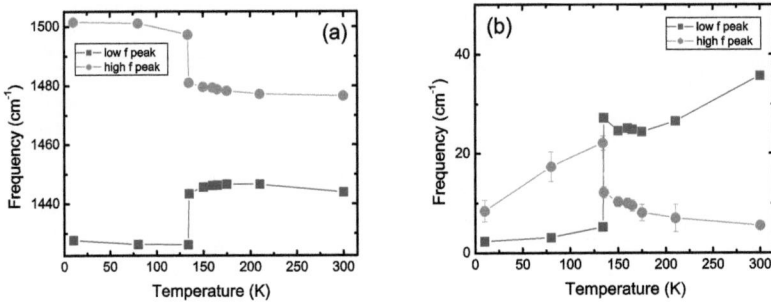

Figure 9.42.: Temperature dependence of (a) position and (b) width of the contributions to the $B_{1u}(\nu_{27})$ mode in $h8$-α-(BEDT-TTF)$_2$I$_3$.

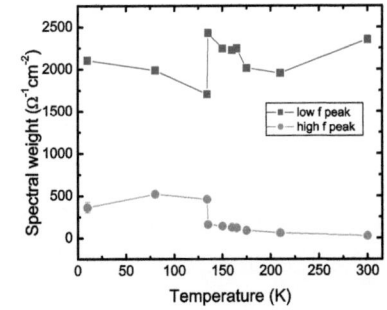

Figure 9.43.: Spectral weight of the three contributions of the $B_{1u}(\nu_{27})$ mode of $h8$-α-(BEDT-TTF)$_2$I$_3$ as function of temperature.

9.2. The metal insulator transition in α-(BEDT-TTF)$_2$I$_3$ and θ-(BEDT-TTF)$_2$RbZn(SCN)$_4$

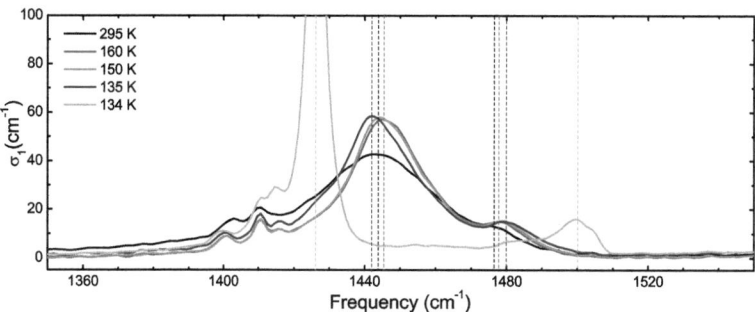

Figure 9.44.: Development of the different parts of the $B_{1u}(\nu_{27})$ mode in $h8$-α-(BEDT-TTF)$_2$I$_3$. Positions of the contributions visualized by dashed lines (see description in text).

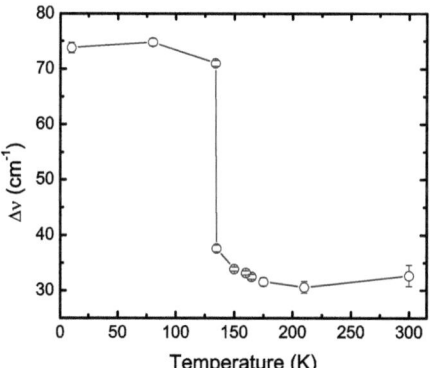

Figure 9.45.: Splitting of the $B_{1u}(\nu_{27})$ peaks in $h8$-α-(BEDT-TTF)$_2$I$_3$ as function of temperature.

151

9. Vibrational spectroscopy

Fig. 9.41 the whole transition traced by the $B_{1u}(\nu_{27})$ mode takes place within a temperature range of 1 K. Therefore the temperature was changed in small steps[1] from the onset of the transition at 135 K (still a broad band) till its end at 134 K (sharp split bands). Already 0.1 K below the onset two broad contributions at 1443 and 1480 cm^{-1} loose weight and the two narrow contributions at 1428 and 1500 cm^{-1} are present. On further cooling the first two peaks loose additional weight on expanse of the latter two. Already 1 K below the onset of the transition the broad peaks are gone. There is no gradual shift of the $B_{1u}(\nu_{27})$ contributions and there is a coexistence of both components in the spectrum at 134.9 K in Fig. 9.41. The amount of charge redistribution changes abrupt at the transition. However, right before the transition the charge redistribution increases as evidenced by the increased splitting between the peaks shown later in Fig. 9.45.

The position and width of the two main contributions is extracted from fits to the spectra over the whole temperature range and presented in Fig. 9.42. The fits are made in the same way as for the compounds presented before including an oscillator describing the low frequency background. Here the small splitting at lowest temperatures into the pairs of peaks (Fig. 9.40) was not taken into account. Especially the (a) peak is difficult to distinguish from the background. But the general behavior of the strong main peak (b) is not affected by that. The splitting of the peaks (Fig. 9.42 (a)) is already present at room temperature. On cooling to 200-180 K both peaks show a slight hardening. On cooling further towards the transition temperature at 135 K the high frequency peak hardens even stronger while the low frequency peak softens: The the splitting between them gradually increases. At the transition itself the splitting between the modes increases jump-like and then stays constant. Below the transition both peaks slightly harden, again. The extracted temperature behavior in the splitting is shown in Fig. 9.45. Also the width of the modes shows a significant temperature dependence: Cooling to 180 K the low frequency peak sharpens while the high frequency peak widens. Cooling further both peaks widen on approaching the transition temperature. Right at the transition the high frequency peak widens step-like while the low frequency peak sharpens. Below the transition both peaks become sharper on decreasing temperature. In terms of spectral weight (Fig. 9.43) above the transition the spectral weight of the peaks increases. (The spectral weight of the low frequency peak is difficult to distinguish from the low frequency background at room temperature, therefore the weight attributed to the peak might be too high). At the transition the the peaks spectral weight changes abrupt. The high frequency peak increases while the low frequency peak deceases.

The behavior of the $B_{1u}(\nu_{27})$ mode and the changes of its single bands is clearly obvious also in the peaks in the spectra. The temperature range down to the transition is given in Fig. 9.44, the splitting between the peaks in Fig. 9.45. A possible description of the behavior is as follows:

- Already at room temperature the $B_{1u}(\nu_{27})$ mode consists of two bands (black lines). The low frequency part located at 1443 cm^{-1} with the greater part of

---

[1] The absolute values of the temperatures are estimated within an error of 0.5 K due to possible temperature gradient between sample on the cold finger of the cryostat and the temperature sensor. However, the relative changes of 0.1 K are not influenced by that.

9.3. Charge Order and charge redistribution for metals, superconductors, and insulators

spectral weight and the high frequency part around 1476 cm^{-1}. The splitting corresponds to a charge disproportionation of $\Delta\rho = 0.24$ e.

- With decreasing temperatures down to approximately 180 K the modes harden by the same amount (red lines). Therefore the splitting stays basically constant and the charge disproportionation remains unchanged.

- Cooling close to the transition temperature of 135 K the two components start to split slightly more (blue lines). The splitting enhances by several wavenumbers but the corresponding charge disproportionation does not exceed 0.29e.

- At the phase transition ($T \approx 135K$) the large splitting occurs. The charge disproportionation increases abruptly as discussed before (cyan lines).

To lower temperatures the charge redistribution increases even more (Fig. 9.39). At lowest temperature the charge redistribution $\Delta\rho$ in $h8$-α-(BEDT-TTF)$_2$I$_3$ is about 0.5-0.7e, depending what charge-density to center-frequency slope 105-140 cm^{-1}/e is to apply [100]. (For the calculated values above always 140 cm^{-1}/e was assumed). That is in agreement with the charge disproportionation of about 0.6e found in Raman measurements [73].

9.3. Charge Order and charge redistribution for metals, superconductors, and insulators

Including the dc properties, three different characteristic regimes can be distinguished by IR vibrational spectroscopy to trace the charge distribution over the molecular sites. The findings are summed up in Fig. 9.46 (a)-(c). First the pure metallic state (a) of equally distributed charges. In the vibrational spectrum a single contribution to the $B_{1u}(\nu_{27})$ mode is present. The presence of electronic correlations leads to the onset of charge fluctuations; this causes a broadening of the single mode. In the proximity of the charge order transition, the presence of fluctuating charge order patterns results in a distinct splitting of the charge sensitive mode, as seen in Fig. (b). The transport properties are still metallic. That is why this state is characterized as charge-ordered metallic state. For the correlated metals investigated here no significant or only a very weak splitting is found. The charge disproportionation is less than 0.1e in the β''-(BEDT-TTF)$_2$SF$_5$CHFSO$_3$. The metallic α-(BEDT-TTF)$_2$TlHg(SCN)$_4$ is higher correlated and reveals a more significant charge redistribution of about 0.1-0.18e. The superconductors, β''-(BEDT-TTF)$_2$SF$_5$CH$_2$CF$_2$SO$_3$ β-(EDT-TTF)$_4$[Hg$_3$I$_8$]$_{(1-x)}$ and α-(BEDT-TTF)$_2$-NH$_4$Hg(SCN)$_4$ show a significant splitting. The corresponding charge redistribution is about 0.2 to 0.25e. That locates them well in the charge-ordered metallic phase. For the insulators (c) a large splitting is found. At the charge order transition the charge redistribution jumps to 0.6-0.7 for θ-(BEDT-TTF)$_2$RbZn(SCN)$_4$ and α-(BEDT-TTF)$_2$I$_3$. Finally one can conclude that the increasing correlations in the organic conductors lead to an increase of charge fluctuation. They form charge-order patterns of which the

9. Vibrational spectroscopy

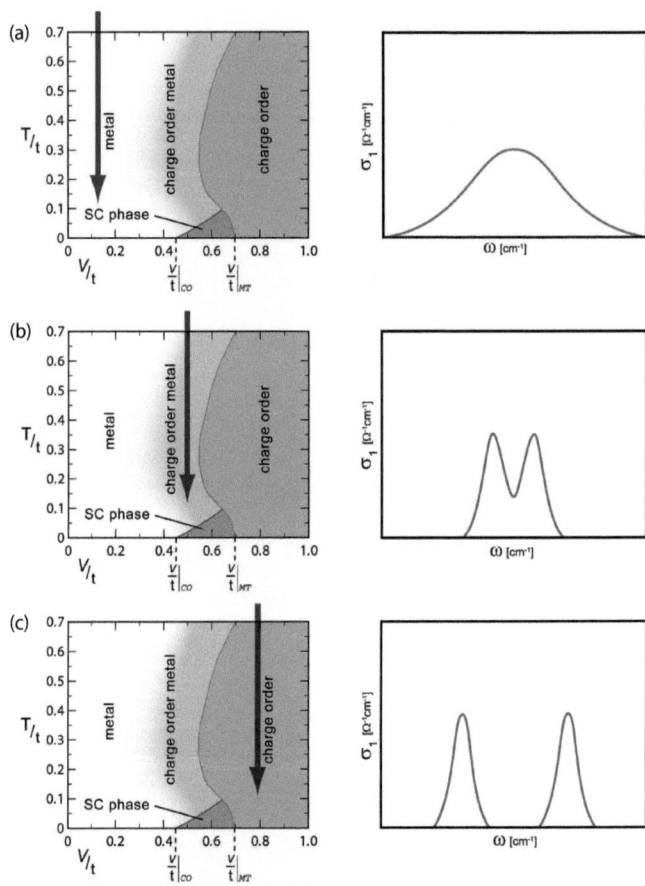

Figure 9.46.: The $B_{1u}(\nu_{27})$ vibrational mode in different correlation regimes. On the left the phase diagram is shown with the position in the (a) metallic, (b) charge ordered metallic, and (c) charge ordered insulating state. The corresponding behavior observed in the $B_{1u}(\nu_{27})$ is sketched on the right. The charge redistribution between the sites leads to the splitting of the mode.

9.3. Charge Order and charge redistribution for metals, superconductors, and insulators

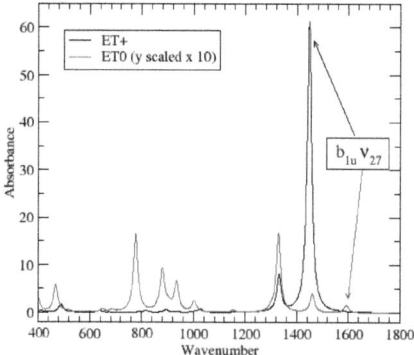

Figure 9.47.: Sample calculation of the vibrational modes in the neutral (ET0, red) and fully charged (ET+, black) BEDT-TTF molecule. The $B_{1u}(\nu_{27})$ vibrational mode has a tremendous increase of absorption when going from the neutral to the fully ionic state. One has to note that the values for the neutral state a scaled by a factor of 10. Calculation from A. Girlando, Parma University.

charge disproportionation is detected in the split contributions to the $B_{1u}(\nu_{27})$ IR and the $A_g(\nu_2)$ Raman modes. The highly e-mv coupled $A_g(\nu_3)$ mode in Raman also evidences the presence of fluctuations instead of a static charge-order. For superconductors the presence of these fluctuations in the charge-ordered metallic state are proven with a charge redistribution of $\Delta\rho \approx 0.2e$. For charge order insulators the charge disproportionation is significantly higher. At the charge order transition the redistribution increases to more than $\Delta\rho \approx 0.5e$. No gradual change of charge redistribution right at the transition is observed. However, for α-(BEDT-TTF)$_2$I$_3$ a slight increase right above the transition is found. That could mean the charge fluctuations increase, and then stable charge order patterns are stabilized at the transition.

The strong change in the charge density is also evident in the strength of the single contributions of the split mode. The experiments suggest that the higher the charge disproportionation the larger is the spectral weight change in the single components. For the small redistributions, as in the $\beta-$, $\beta''-$series or the metallic and superconducting α-systems, the single contributions are of approximately equal strength. Going to the charge-ordered insulators, θ-(BEDT-TTF)$_2$RbZn(SCN)$_4$ and α-(BEDT-TTF)$_2$I$_3$, one contribution increases enormous while the other contributions are very small. The comparison of the transition in these two compounds point out that this is mainly an effect of the charge on the molecule and not a reason due to structural changes. Only the θ-(BEDT-TTF)$_2$RbZn(SCN)$_4$ exhibits a structural transition at the metal-insulator transition. The α-(BEDT-TTF)$_2$I$_3$ undergoes the charge order transition without struc-

9. Vibrational spectroscopy

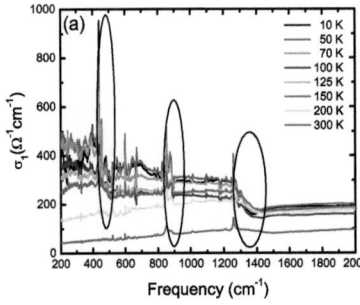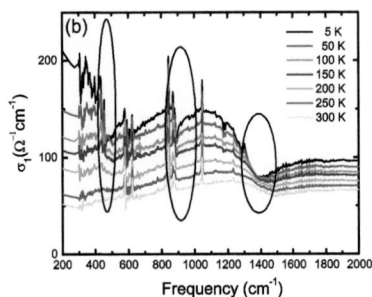

Figure 9.48.: The highly e-mv coupled vibrations in (a) β''-(BEDT-TTF)$_2$SF$_5$CH$_2$CF$_2$SO$_3$ and (b) β''-(BEDT-TTF)$_2$SF$_5$CHFSO$_3$. Most prominent are the $A_g(\nu_{10})$ at 450 cm^{-1}, the $B_{3g}(\nu_{60})$ (if assigned as C_{2v} mode as in [66]) at 890 cm^{-1}, and the $A_{1g}(\nu_2)$ vibration at 1450 cm^{-1}. The peaks below the Fano shaped steps are at 440 cm^{-1} the $A_g(\nu_9)$, at 850 cm^{-1} a SF$_5$ anion mode, and at 1300 the $A_{1g}(\nu_3)$ vibrational mode.

tural changes. But in both systems the splitting of the charge sensitive $B_{1u}(\nu_{27})$ shows a similar behavior with the large changes in the spectral weight of the contributions. This is also strengthened by calculations by A. Girlando on the absorption of the BEDT-TTF molecule in its neutral and fully ionic state presented in Fig. 9.47. A tremendous change in the vibrational spectrum takes place. The absorption of the $B_{1u}(\nu_{27})$ increases by about two orders of magnitude and shifts to lower frequencies. In that sense static charge-order leading to high charges on the molecular sites can lead to an enhancement of the absorption band.

9.4. Properties of the in-plane emv-coupled modes

To be able to fit the conductivity spectra of the organic conductors in the in-plane directions successfully, one needs to use Fano line shapes to describe some of the vibrational features. These modes correspond to fully symmetric vibrations and therefore, missing a dipole moment, should not be seen in the IR spectra. The reason that they are IR active is a symmetry break due to e-mv coupling. They are coupled to the electronic background as described in Sec. 3.3.2 and therefore gain a dipole moment. One of the possibilities to describe their spectra is Fano coupling [113], where the single energy level of the phonon is coupled to the broad electronic background of the band. The most prominent modes used to fit the spectra in the organic superconductor β''-(BEDT-

9.4. Properties of the in-plane emv-coupled modes

TTF)$_2$SF$_5$CH$_2$CF$_2$SO$_3$ and the isostructural metal β''-(BEDT-TTF)$_2$SF$_5$CHFSO$_3$ are shown in Fig. 9.48. Important to notice is that the Fano profile of the mode is the step in the conductivity spectrum. The peaks below these steps are additional modes. In the β''-(BEDT-TTF)$_2$SF$_5$CHFSO$_3$ the lowest mode at 450 cm^{-1} was not fitted, since it is at the border of the measured frequency range. The step like Fano shape shows the strong coupling of the modes to the electronic background. As discussed in [66] the main features increase below 60 K where the electronic background of the MIR band starts to saturate or even decrease. This could be a hint to a small dimerization taking place within the lattice, however that is in disagreement with the slightly reduced dimerization in the transfer integrals found in the x-ray studies (c.f. Sec. 4.1) [116]. The low frequency modes at 318 and 240 cm^{-1} discussed in [66] are assumed here to be coupled to the low frequency charge fluctuation band that shows up. This could be a fingerprint to a polaronic enhancement of the band [178]. The modes are e-mv coupled to it and they grow in parallel.

A successful fit of the vibrational features enables to see and separate the underlying bands and enables the description of the charge fluctuation band as additional contribution to the MIR band in the subsequent description of the in-plane optical response.

10. Broad band optical measurements

The optical properties of the organic superconductors β''-(BEDT-TTF)$_2$-SF$_5$CH$_2$CF$_2$SO$_3$ and an isostructural metal β''-(BEDT-TTF)$_2$SF$_5$CHFSO$_3$ are investigated in a broad frequency (8-8000 cm^{-1}) and temperature (1.8-300 K) range. Here we focus on the in-plane properties. The combination of FTIR and THz spectroscopy reveals the optical response of the system showing traces of quasi free charge carriers as well as localization effects. The influence of charge fluctuations becomes manifest in an extended Drude analysis and in an electronic charge fluctuation band in the low frequency optical conductivity. In the THz range the superconducting gap was investigated. Further the spectra of the β-(EDT-TTF)$_4$[Hg$_3$I$_8$]$_{(1-x)}$ superconductor is presented in the accessible frequency range to compare the dynamical properties.

10.1. The in plane response of the β''-(BEDT-TTF)$_2$SF$_5$CH$_2$CF$_2$SO$_3$ superconductor and the isostructural β''-(BEDT-TTF)$_2$SF$_5$CHFSO$_3$ metal

To obtain the broadband optical response of the organic superconductor β''-(BEDT-TTF)$_2$SF$_5$CH$_2$CF$_2$SO$_3$ and its isostructural sister compound, the organic metal β''-(BEDT-TTF)$_2$SF$_5$CHFSO$_3$, their optical reflectivities are measured. The reflectivity of the β''-(BEDT-TTF)$_2$SF$_5$CH$_2$CF$_2$SO$_3$ along the a- and b-axis of the system in the whole measured frequency range is shown in Fig. 10.1. The inset shows the orientation of the crystal axes within the conducting plane. The c-axis is oriented out-of-plane. Measurements were taken along the highest reflecting b-direction and perpendicular to it what approximately corresponds to the a-direction. For the in-plane properties at room temperature a frequency range from 8-25000 cm^{-1} was measured for the β''-(BEDT-TTF)$_2$-SF$_5$CH$_2$CF$_2$SO$_3$ (Fig. 10.1). Crystal size effects limit the low frequencies to 20 cm^{-1} in the high conducting b-direction. In the β''-(BEDT-TTF)$_2$SF$_5$CHFSO$_3$ the low frequency limit is 300 cm^{-1} due to the small spot size in the IR microscope. The reflectivity shows a minimum around 7000 cm^{-1} marking the upper part of the conduction band. Above that frequency, an approximately polarization independent and broad band around 30000 cm^{-1} sets in. This band is due to intramolecular electronic transitions [66]. For these transitions no temperature dependence is expected. Also no significant changes are found in temperature dependent NIR reflectivity measurements up to 12000 cm^{-1}. The onset of the intramolecular transitions is presented only for room temperature here. But it is taken into account as high frequency extrapolation in all fits for the temperature dependent spectra. It is basically the same for all BEDT-TTF charge transfer salts of

10. Broad band optical measurements

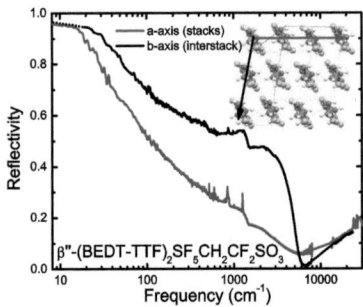

Figure 10.1.: Broadband room temperature reflectivity measured in β''-(BEDT-TTF)$_2$-SF$_5$CH$_2$CF$_2$SO$_3$ including the onset of the high frequency band of the interband transitions. The projection of the crystal structure to the conducting plane and the crystallographic axes are given as inset.

similar structure. Therefore the high frequency extrapolation for the β''-(BEDT-TTF)$_2$-SF$_5$CH$_2$CF$_2$SO$_3$ is also used in the isostructural β''-(BEDT-TTF)$_2$SF$_5$CHFSO$_3$.

The reflectivity and optical conductivity of these systems at room temperature along the principle directions is shown in Fig. 10.2 (a) and (b). As already seen in the reflectivity data, the conduction band spans up to frequencies of approximately 7000 cm^{-1}. It describes the optical response of the conducting charge carriers. A low frequency Drude peak shows the coherent transport while the MIR band describes the localization effects under the influence of the electron-electron repulsion.

Both systems show a significant in-plane out-of-plane anisotropy of $\sigma_{ab}/\sigma_c \approx 100$ while the in-plane anisotropy $\sigma_a/\sigma_b \approx 5$ is rather small. This, in addition to the transport characterization (Sec. 8), shows the pronounced two-dimensionality of the systems. However, the in-plane anisotropy is significant and larger than for other BEDT-TTF based systems [30] pronouncing some one-dimensional character of the system. The increasing reflectivity to lowest frequencies in Fig. 10.2 describes the metallic behavior in both the superconductor β''-(BEDT-TTF)$_2$SF$_5$CH$_2$CF$_2$SO$_3$ and the metal β''-(BEDT-TTF)$_2$-SF$_5$CHFSO$_3$.

The MIR band in both compounds is located around 1500-2000 cm^{-1}. For the polarization along the interstack b-direction it emerges in reflectivity up to a drop at about 6000 cm^{-1}. Along the stacks, in the a-direction, an over-damped behavior of the reflectivity is observed. The conductivity minimum above the MIR band is around 7000 cm^{-1} along both in-plane polarizations and in both compounds. It sets the high frequency limit of the conduction band.

10.1. The in plane response of the β''-$(BEDT\text{-}TTF)_2SF_5CH_2CF_2SO_3$ superconductor and the isostructural β''-$(BEDT\text{-}TTF)_2SF_5CHFSO_3$ metal

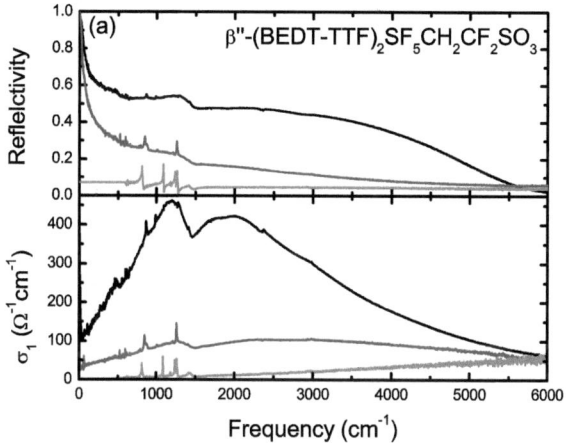

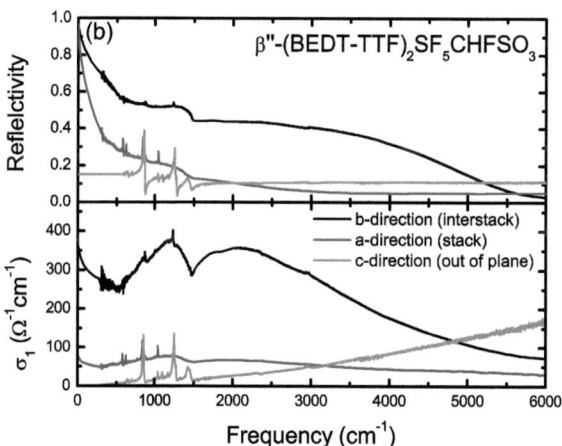

Figure 10.2.: In plane reflectivity and optical conductivity of (a) β''-$(BEDT\text{-}TTF)_2$-$SF_5CH_2CF_2SO_3$ and (b) β''-$(BEDT\text{-}TTF)_2SF_5CHFSO_3$ along the principal axes at room temperature.

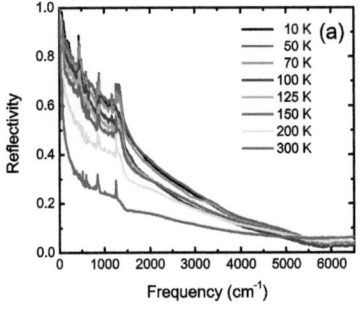
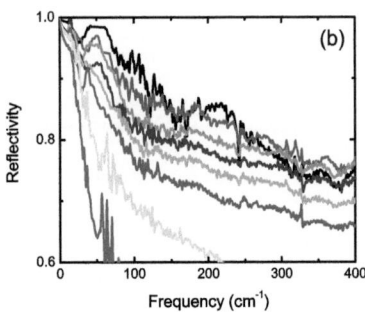

Figure 10.3.: (a) Temperature dependence of the in-plane reflectivity (E||a) of the β''-(BEDT-TTF)$_2$SF$_5$CH$_2$CF$_2$SO$_3$ along the stacks. Fig. (b) shows the behavior in the low frequency regime.

The high frequency contribution above that band is important in terms of spectral weight conservation. On decreasing temperature the organic crystals shrink, as e.g. seen in the expansion coefficient [130] [1]. This leads to an altered molecular orbital overlap between the sites and to an increased bandwidth of the conduction band. The transfer of spectral weight that goes along with that increase comes from spectral weight changes in the intramolecular transition band in order to conserve the total spectral weight.

10.2. Temperature dependence of the in-plane optical response

The focus here is on the charge carrier dynamics within the conduction band. The measured temperature dependent conductivity along the in-plane a- and b-axes are presented and discussed in the following. The findings are also put into relation to the vibrational properties from Sec. 9 with view on charge disproportionation and fluctuation.

10.2.1. β''-(BEDT-TTF)$_2$SF$_5$CH$_2$CF$_2$SO$_3$

Temperature dependent reflectivity of β''-(BEDT-TTF)$_2$SF$_5$CH$_2$CF$_2$SO$_3$ for the light polarized along and perpendicular to the stacks is presented in Figs. 10.3 and 10.4. The respective conductivities are plotted in Fig. 10.6 (a) and (b). The system shows metallic

---

[1] Another experimental fact of the shrinking of the crystal comes from the gold evaporation experiment. There a golden film is evaporated on the crystal surface. Due to the non-uniform and significantly different shrinking of the crystal the film peals off the surface on cooling.

10.2. Temperature dependence of the in-plane optical response

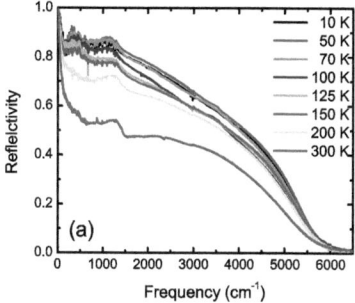
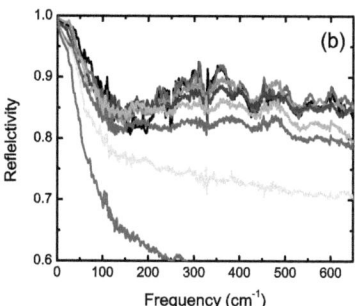

Figure 10.4.: (a) Temperature dependence of the in plane reflectivity of the β''-(BEDT-TTF)$_2$SF$_5$CH$_2$CF$_2$SO$_3$ perpendicular to the stacks (E∥b). Fig. (b) shows the behavior in the low frequency regime.

properties. It becomes more reflecting and conducting with decreasing temperature. In conductivity the main feature is the broad MIR band dominating at room temperature. On decreasing temperature spectral weight is transfered to lower frequencies.
At 150 K a feature starts to appear at the low frequency edge of the MIR band. The reflectivity starts to flatten around the 200 and 300 cm^{-1} region. At temperatures below 150 K even a gap-like drop in reflectivity is observed, especially in the high reflecting b-direction (perpendicular to the stacks). In conductivity that leads to the appearance of a low frequency feature which grows on further cooling. The detailed evolution of this feature is shown as insets in the Fig. 10.3 and Fig. 10.4 for the reflectivity and in Fig. 10.7 for the optical conductivity (a) along and (b) perpendicular to the stacks. In the a-direction (along the stacks) below 300 cm^{-1} the slope of the reflectivity starts to change with an increasing effect for decreasing temperatures. Below 150 K a small hump around 200 cm^{-1} builds up.
On lowest temperatures the influence of e-mv coupled phonons and vibrations can be seen at 120, 170, and 240 cm^{-1}. In the b-direction the effects are even more pronounced. The flattening of reflectivity takes place below 500 cm^{-1} and a maximum appears around 300 cm^{-1}. It is followed by a drop to a minimum around 150 cm^{-1}. Low frequency phonons and vibrations are present at 110, 250, 320, and 420 cm^{-1}. In both directions the hump and the enhancement of reflectivity turns into a low frequency band in conductivity for temperatures below 150 K. The band is located around 200 and 300 cm^{-1} for the stack direction and the interstack direction respectively. That happens right at the temperature where first traces of charge fluctuations are observed in the transport properties and IR vibrational measurements.
The flattening and drop of reflectivity to low frequencies leads to the opening of a pseudo-

10. Broad band optical measurements

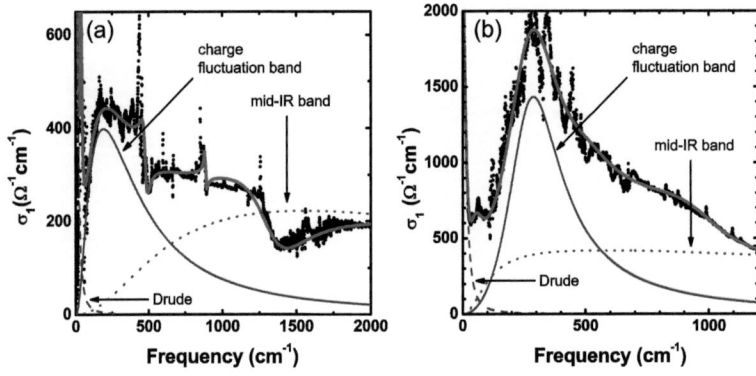

Figure 10.5.: Drude Lorentz fit (red) to the in-plane optical conductivity (black circles) of the β''-(BEDT-TTF)$_2$SF$_5$CH$_2$CF$_2$SO$_3$ along (a) and perpendicular (b) to the stacks. The main contributions (blue) due to MIR band (dotted line, the full band see Fig. 10.6), charge fluctuation band (full line) and Drude response (dashed line) are shown. The spectral weight shift from high to low frequencies of the bands compared to the fit is due to the Fano shape used to describe the e-mv coupled lattice vibrations (c.f. Sec. 9.4).

gap below the MIR band. Below 150 cm^{-1} the reflectivity shows an up-turn, again. That corresponds to a very low frequency Drude response in conductivity that fills in the gap. Parallel to the stacks an additional feature is seen in reflectivity around 50 cm^{-1} which arises for temperatures below 150 K. In conductivity that leads to the onset of a mode around 40 cm^{-1} which grows with low temperatures. It is interesting to notice that the reflectivity enhancement has a maximum around 50 K while the lower reflectivity gap still opens more on decreasing temperature. In conductivity it is seen that the corresponding band in the 200 to 300 cm^{-1} region also has its maximum around 50 K while the 40 cm^{-1} mode and the Drude response still increase with lower temperatures.

Summarizing, the spectrum contains three main contributions as seen from a Drude Lorentz fit to the conductivity along both directions (Fig. 10.5). The strong MIR band centered around 700 cm^{-1} with an optical gap below it. At lowest frequencies there is a very narrow Drude response of several wavenumbers width, only. For low but finite frequencies a band around 200-300 cm^{-1} exists. It is assigned as charge fluctuation band as motivated by the discussion that follows. Parallel to the stacks also another feature at about 40 cm^{-1}, not shown in the fit, can be identified as strong low frequency phonon

10.2. Temperature dependence of the in-plane optical response

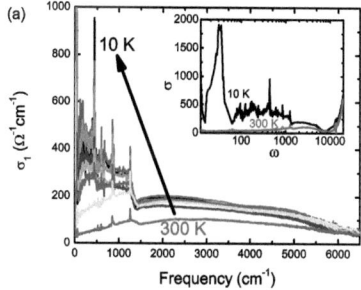 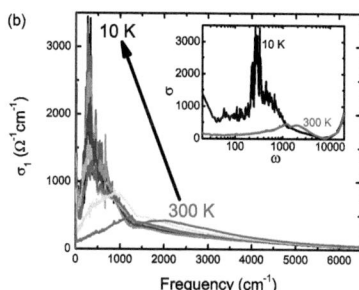

Figure 10.6.: Temperature dependent in-plane optical conductivity of the β''-(BEDT-TTF)$_2$SF$_5$CH$_2$CF$_2$SO$_3$ (a) along and (b) perpendicular to the stacks. The insets show the 300 and 10 K data in a logarithmic scale.

which acts as collective charge order excitation.

The charge transfer band and Drude response

The optical response of the β''-(BEDT-TTF)$_2$SF$_5$CH$_2$CF$_2$SO$_3$ at high temperatures is dominated by the broad MIR band (Fig. 10.6). With decreasing temperature a red-shift of spectral weight is observed (Fig. 10.8 (a)). Below the MIR band an optical pseudo-gap is present. The band expresses the presence of on-site and inter-site electronic correlations in the system as described in Sec. 3. According to that, the position of the MIR band corresponds approximately to $3V$ and can be identified as Hubbard-like charge transfer band. The increase of the spectral weight of it is due to the increasing influence of the correlations which lead to stronger localization effects.
Notable, a small drop in spectral weight takes place at 150-125 K (Fig. 10.8 (b)). That is right the onset temperature where transport and vibrational analysis traces of charge fluctuations in the metallic state. The red-shift of the band is caused by the temperature dependent shrinking of the crystal. It is anisotropic: Along the stacks the system shrinks gradually while perpendicular to the stacks a strong shrinking at high temperatures appears which then stays basically constant. Due to the shrinking lattice constants the overlap of the molecular sites increases and therefore changes the effective correlation V/t in the system. These basically determine the position of the MIR band (Sec. 3).
For temperatures below 150 K an electronic band starts to develop within the gap as seen in the zoomed in spectra in Fig. 10.7 and which is clearly disentangled in the fit (Fig. 10.5). The band is located around 200 cm^{-1} and 300 cm^{-1} for the direction along and perpendicular to the stacks respectively. This band is associated with the interaction of charge carriers with the charge fluctuations in the system and will be addressed as

10. Broad band optical measurements

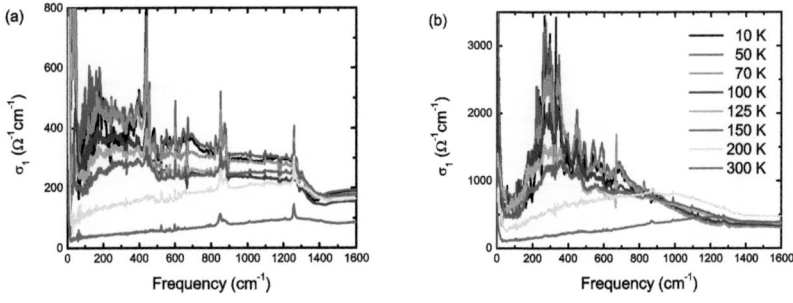

Figure 10.7.: Low frequency zoom in into the temperature dependent in-plane optical conductivity of the β''-(BEDT-TTF)$_2$SF$_5$CH$_2$CF$_2$SO$_3$ (a) along and (b) perpendicular to the stacks. Below the MIR band the charge fluctuation band and the zero frequency Drude response is visible.

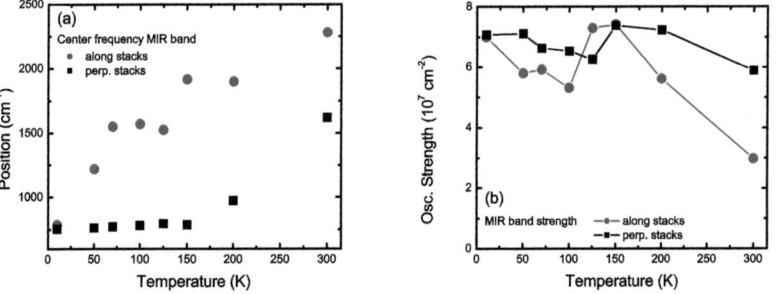

Figure 10.8.: (a) Position and (b) spectral weight (represented by its oscillator strength) of the MIR band in β''-(BEDT-TTF)$_2$SF$_5$CH$_2$CF$_2$SO$_3$. The direction along the stacks is given as red dots and the values perpendicular to the stacks as black squares. The lines are drawn to visualize the drop of oscillator strength with onset of the charge fluctuations at around 150 K.

10.2. Temperature dependence of the in-plane optical response

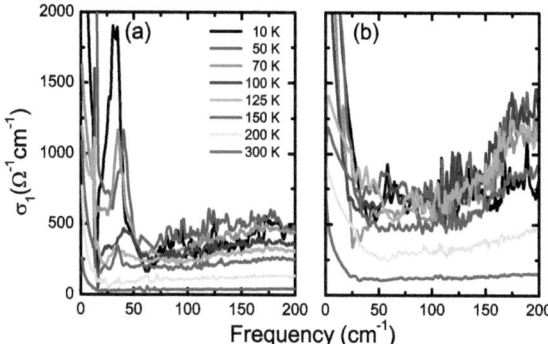

Figure 10.9.: Temperature dependence of the Drude response in β''-(BEDT-TTF)$_2$-SF$_5$CH$_2$CF$_2$SO$_3$ (a) along and (b) perpendicular to the stacks.

charge fluctuation band later in Sec. 10.2.1. The appearance of this band together with the drop of spectral weight in the MIR band suggests a transfer of spectral weight into it.

For all temperatures, and with a width of approximately 10 cm^{-1}, the onset of the Drude response is observed at lowest frequencies. The upturn of conductivity to the Drude peak is clearly seen along both directions (Fig 10.9). A small response is already present at room temperature and increases with decreasing temperature.

The spectrum along the stacks (Fig. 10.7(a) and 10.9 (a))shows a strong feature around 30-40 cm^{-1} which increases for low temperatures. It turns out to be a collective charge order excitation in terms of a lattice phonon as discussed in Sec. 10.2.1. Perpendicular to the stacks this feature is not observed.

The spectral weight of the Drude response shows an increases with decreasing temperature along both directions. That is seen in the increasing plasma frequency which is extracted from the Drude Lorentz fit (Fig. 10.10). The existence of the Drude peak directly reveals the metallic response of the system. This is fully in line with the metallic transport properties. However, for all temperatures the spectral weight of the Drude response stays very small. The exact Drude spectral weight is difficult to integrate since the high frequency part of the peak is measured and the low frequency part is influenced by the extrapolation. Nevertheless, as shown by the red dots in Fig. 10.11 at 10 K only about 3% of the conduction bands spectral weight is in the Drude response. That means that only a small fraction of charge carriers contributes to the coherent transport. For higher temperatures an approximately linear decrease of the relative spectral weight is

167

10. Broad band optical measurements

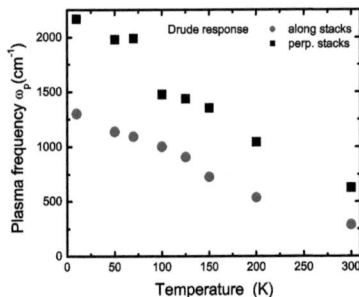

Figure 10.10.: Temperature dependence of the plasma frequency extracted from the Drude Lorentz fit perpendicular (black square) and along the stacks (red dots) in β''-(BEDT-TTF)$_2$SF$_5$CH$_2$CF$_2$SO$_3$.

observed. For 150 K and above the Drude contribution even drops well below 1% of the conducting bands spectral weight. That shows the collapse of the coherent quasiparticles with higher temperatures. That is a typical behavior known for bad metals.

The extended Drude analysis: Interaction with charge fluctuations

To fit the low frequency part of the spectrum it turns out that a simple Drude response is not sufficient. The deviation from a simple Drude behavior is due to additional interactions of the charge carriers. To gain insight into that interaction an extended Drude analysis is performed. Therefore the frequency dependent scattering rate and effective mass are calculated as described in Sec. 2.1.2. Usually the plasma frequency extracted from the spectral weight of the Drude peak enters the extended Drude formulas. But the Drude response is so narrow that only the high frequency part of the Drude could be measured and already overlaps strongly with the gap features. The plasma frequency extracted from the integral over the Drude response is very inaccurate then. One reason is that the low frequency part of the Drude spectral weight comes from the Hagen Rubens extrapolation. Further and more significant is that the separating of the electronic features from the pure Drude response can not be done perfectly. Therefore the obtained plasma frequency from the Drude Lorentz fits (Fig. 10.10) was used within the extended Drude analysis.

The results are given in Fig. 10.12 for the spectra perpendicular to the stacks. Figure (a) shows the Drude response in the conductivity. The increasing Drude response is evident. Also the reentrant like behavior into the more metallic state for low temperatures is obvious. Spectral weight is transfered from the charge fluctuation band to the Drude

10.2. Temperature dependence of the in-plane optical response

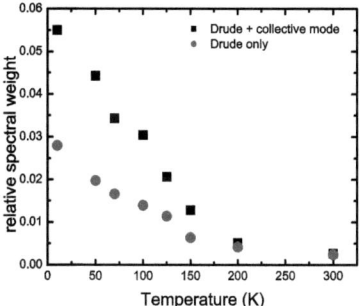

Figure 10.11.: Relative spectral weight of the Drude response compared to the conduction band total spectral weight of β''-(BEDT-TTF)$_2$SF$_5$CH$_2$CF$_2$SO$_3$ along the stacks. The red dots represent the relative spectral weight of the free Drude response, only. The black squares the combined relative spectral weight of the Drude response and the collective low frequency mode coupled to a lattice phonon.

below 100 K. The frequency dependent scattering rate and the respective effective mass are given in (b) and (c). The region with positive effective mass marks the range where the interpretation within the extended Drude picture is meaningful. Only for that region the response of the quasi-free charge carriers is taken into account. At room temperature it spreads up to about 60 cm^{-1}. With decreasing temperature the range widens up to 100 cm^{-1}. That shows the overall increasing Drude response. The same holds for the region where the scattering rate increases with increasing frequency. For high temperatures the results are washed out with thermal fluctuations so that for an interpretation the low temperature data is of importance. That was already seen in the small fraction of the Drude spectral weight compared to the conducting band. At high temperatures the quasiparticle picture is not usable anymore and the extended Drude parameters cannot be interpreted. Therefore at low temperatures the scattering rate and effective mass at 10 and 30 K are given separately in Fig. 10.13. The linear frequency dependence of the scattering rate in the range from 30-100 cm^{-1} suggest an interaction of the charge carriers with bosonic excitations. Here it is interaction with the charge fluctuations in the system. Below 30 cm^{-1} it cannot be judged whether the frequency dependence shows a quadratic behavior as expected for a Fermi-liquid, or a frequency independence as for a simple Drude response. The increase of effective mass to lower frequencies at the same time represents the excitation cloud of the charge fluctuations which is carried by the charge carriers. The absolute values of the low frequency effective mass of about $4m_e$

10. Broad band optical measurements

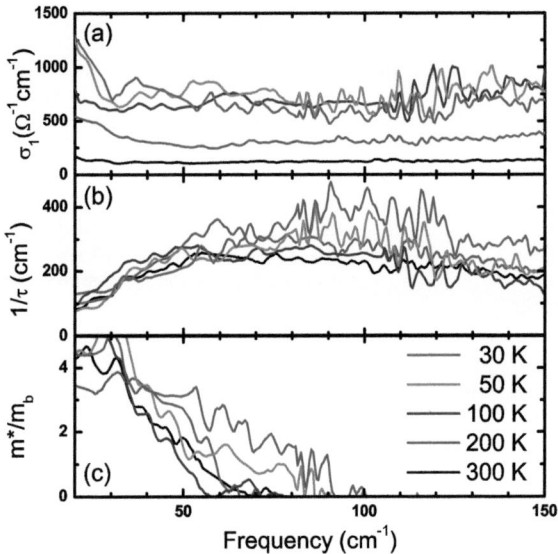

Figure 10.12.: Temperature dependent extended Drude analysis for β''-(BEDT-TTF)$_2$-SF$_5$CH$_2$CF$_2$SO$_3$: Frequency dependent (a) conductivity, (b) scattering rate, and (c) effective mass perpendicular to the stacks.

10.2. Temperature dependence of the in-plane optical response

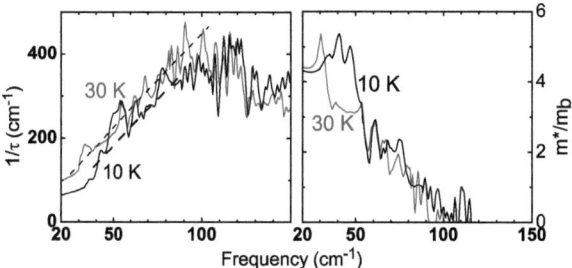

Figure 10.13.: Frequency dependent (a) scattering rate and (b) effective mass perpendicular to the stacks extracted from β''-(BEDT-TTF)$_2$SF$_5$CH$_2$CF$_2$SO$_3$ using extended Drude analysis at 10 and 30 K.

correspond well to the ones which are extracted as cyclotron effective mass or Shubnikov-de Haas effective mass [120, 123].

The extended Drude analysis in the direction along the stack is shown in Fig. 10.14. In the range of (a) the Drude response in optical conductivity, (b) the scattering rate increases up to around 50 cm^{-1} for high temperatures. That is the same behavior as observed for the perpendicular direction. The slope and the absolute values are the same along both directions showing the anisotropy in the coherent response. The effective mass enhancement (c) is also similar along both directions: At frequencies above 50 cm^{-1} the effective mass become negative at the breakdown of the extended Drude description. That is right the frequency where the scattering rate starts to flatten. To lower frequencies the linear increase of the effective mass shows the dressing of the electrons. Absolute values are slightly larger along the stack direction being around 4-5 m_e at lowest frequencies. At lower temperatures also a linear increase of the scattering rate with frequency is observed but already above 30-40 cm^{-1} the influence of a low frequency phonon feature overlaps with the pure coherent contribution and does not allow an interpretation of scattering rate and effective mass of the quasi-particles.

The low frequency lattice phonon: Collective CO excitation

As stated above, at temperatures below the onset of charge fluctuations a strong feature around 30-50 cm^{-1} perpendicular to the stacks appears in the spectrum. Its spectral weight grows on decreasing temperature (Fig. 10.16). This feature is interpreted as lattice phonon. Its position and width are given in Fig. 10.17. The parameters are taken from the combined reflectivity-conductivity fit performed to describe the spectra. It is

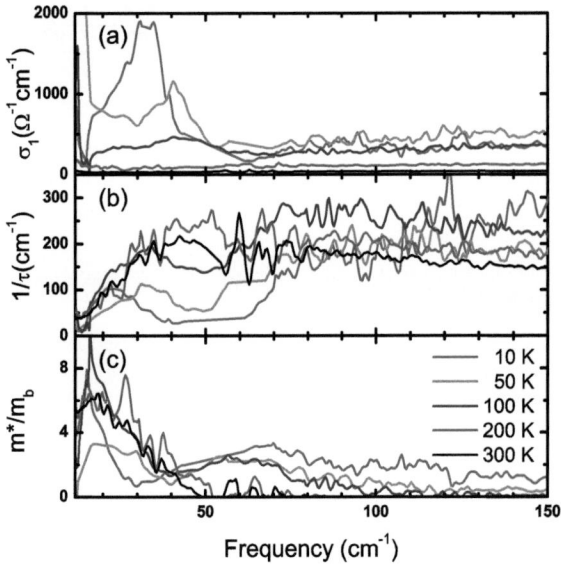

Figure 10.14.: Temperature dependent extended Drude analysis for β''-(BEDT-TTF)$_2$-SF$_5$CH$_2$CF$_2$SO$_3$: Frequency dependent (a) conductivity, (b) scattering rate, and (c) effective mass along the stacks.

10.2. Temperature dependence of the in-plane optical response

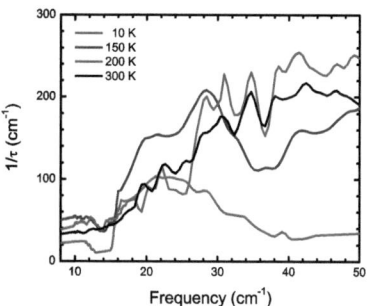

Figure 10.15.: Low frequency zoom in to the frequency dependent scattering rate along the stacks extracted from β''-(BEDT-TTF)$_2$SF$_5$CH$_2$CF$_2$SO$_3$ using extended Drude analysis at several temperatures.

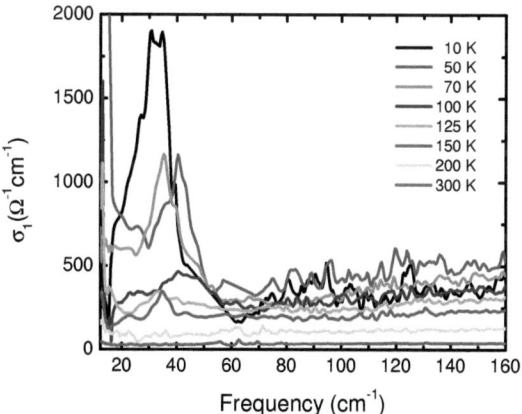

Figure 10.16.: Collective charge order excitation via emv-coupled lattice mode measured along the stacks.

173

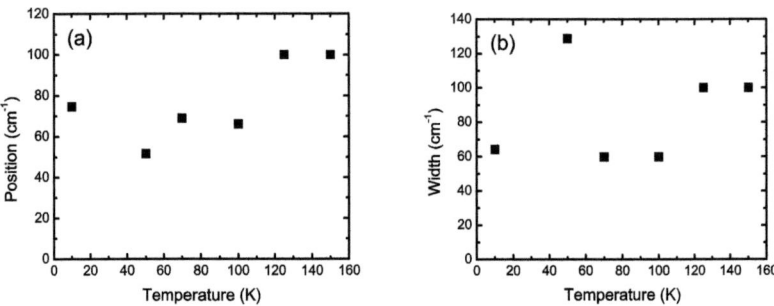

Figure 10.17.: Temperature dependence of the (a) position and (b) width of the low frequency mode along the stacks in β''-(BEDT-TTF)$_2$SF$_5$CH$_2$CF$_2$SO$_3$.

difficult to extract exact values at low frequencies but tendencies of the mode development are possible to see. It shows a slight softening on decreasing temperature while its width stays around 100 cm^{-1}. In general, at these low frequencies lattice modes are expected and found experimentally for similar materials [61, 92, 98, 179–184]. The spectral weight of the mode is plotted in Fig. 10.18 represented by the black squares. It increases in parallel with the electronic background given by the Drude spectral weight shown by the red open circles. Such a strong increase of a phonon is unexpected. The tremendous enhancement on cooling could be explained in terms of an collective excitation of the charge order in the system. This can be motivated by the the extended Drude analysis above (Sec.10.2.1). This already revealed an interaction of the electronic system with a bosonic excitation. Thus we could refer to the mode as collective excitation of the charge order in the system.

The picture is as follows: Charge order fluctuations are present in the system as an unequal charge redistribution between the lattice sites as proven by vibrational spectroscopy. A lattice phonon in that case excites the differently charged molecular sites against each other and in that sense acts as a collective excitation of the charge order. The mode is highly coupled to the electronic system. That explains the strong enhancement of the phonon which goes in parallel with the electronic background. The increase of the mode with decreasing temperature nicely shows that the charge order in the system is strongly enhanced at lower temperatures resulting in a higher probability of interactions between them and the charge carriers.

As summarized in Sec. 3 in the picture of the charge fluctuation model there is also a collective charge excitation driving the charge order transition. It is described as a collective mode in the charge correlation function representing the charge fluctuation in

10.2. Temperature dependence of the in-plane optical response

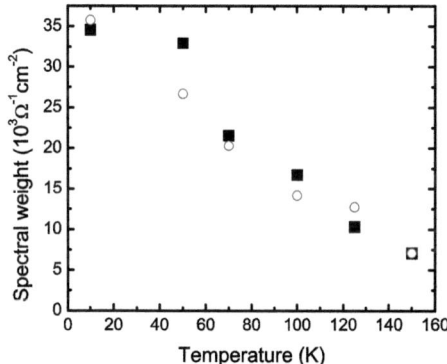

Figure 10.18.: Spectral weight of the low frequency mode (black squares) in β''-(BEDT-TTF)$_2$SF$_5$CH$_2$CF$_2$SO$_3$. The mode increases in parallel to the electronic background represented by the Drude spectral weight given the red open circles.

10. Broad band optical measurements

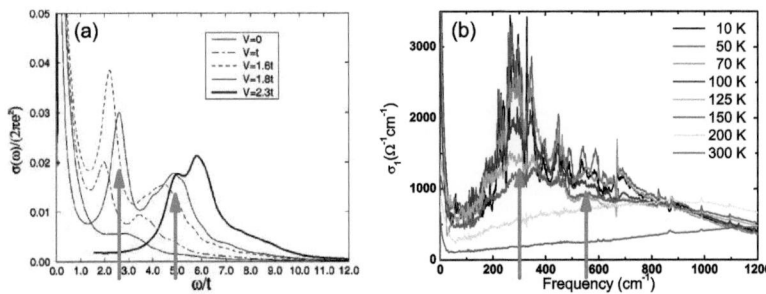

Figure 10.19.: The charge fluctuation band in β''-(BEDT-TTF)$_2$SF$_5$CH$_2$CF$_2$SO$_3$. (a) Calculated spectrum with charge fluctuation band appearing around 2t. From [18]. (b) Temperature dependence for the direction perpendicular to the stacks. The arrows mark the position of the charge fluctuation band and the MIR band at low temperatures.

the system. A direct excitation of that mode is not possible since it is located at (π, π) while light can couple to excitations with $\Delta k = 0$, only. The charge order coupled phonon mode here could be seen then as an equivalent charge excitation mode at (0,0).

In terms of spectral weight, the collective modes spectral weight could be added to the Drude spectral weight as incoherent contribution of a finite frequency quasi-Drude peak. This combined relative spectral weight compared to the conduction band total spectral weight is shown in Fig. 10.11 by the black squares. But also here at 10 K not more than 6% of relative spectral weight is found in the conducting charge carriers. As for the Drude spectral weight, with increasing temperature this combined spectral weight decreases linearly relative to the total band spectral weight. Above 150 K also a drop to less than 1% is observed, too. There the collective mode is absent and the relative spectral weight is basically the pure Drude contribution.

The charge fluctuation band

The band assigned as charge fluctuation band arises within the gap at around 200 and 300 cm^{-1} in the spectra along and perpendicular to the stacks respectively. It is well seen in Fig. 10.5 (a) and (b) where the contribution has been disentangled by the fits. Its temperature dependence is once again shown in Fig. 10.19(b) for the interstack direction. It starts to appear below 200 K (Fig. 10.19 (b), or low frequency part of Fig. 10.6 and Fig. 10.7) together with the onset of charge fluctuations in the system and grows and

10.2. Temperature dependence of the in-plane optical response

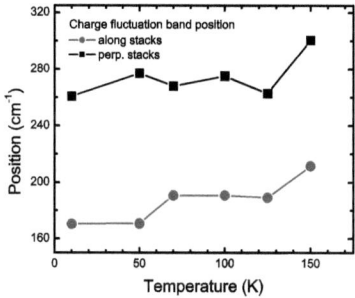 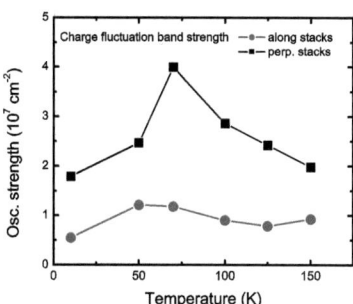

Figure 10.20.: (a) Position and (b) spectral weight (represented by its oscillator strength) of the charge fluctuation band in β''-(BEDT-TTF)$_2$SF$_5$CH$_2$CF$_2$SO$_3$. The direction along the stacks is given as red dots and the values perpendicular to the stacks as black squares.

softens with decreasing temperature. The extracted fit parameters are given in Fig. 10.20. Its maximum intensity is reached at about 70-50 K. Then the reentrant behavior to a more metallic state is observed: The charge fluctuation band decreases again and shifts spectral weight to the Drude part. The band itself can be be described to originate from the transition from the incoherent band to the quasi particle peak in the density of states (sketched in Sec. 3.2.3 and developed theoretically in [18]). The main contributions in the experimental spectrum are in perfect agreement with the contribution in the calculated conductivity spectrum for T=0 (Fig. 10.19 (a), Ref. [18]). It shows the charge fluctuation band as low frequency electronic band around $2t$ in between the Drude response and the MIR charge transfer band. They are marked by arrows in the graph.

A. Greco calculated the temperature dependence of the low frequency response in the framework of charge fluctuations for a comparison with the experiment. The optical conductivity was calculated in the strong coupling limit ($U \rightarrow \infty$) of the extended Hubbard model where a nearest-neighbors Coulomb repulsion V was included [17, 18]. The optical conductivity was computed using the extended Drude formula after the calculation of the self-energy [18]. The phase diagram of the model in the $V - T$ plane (see also Ref. [17]) is shown in the inset of Fig. 10.21. The phase boundary is showing the point where the charge order instability occurs and is calculated as the point where the charge susceptibility diverges. The optical conductivity was calculated for several temperatures along the arrow, on the metallic side but near the charge order. With decreasing temperature, a low energy picture emerges when the system approaches the back-bending of the phase diagram.

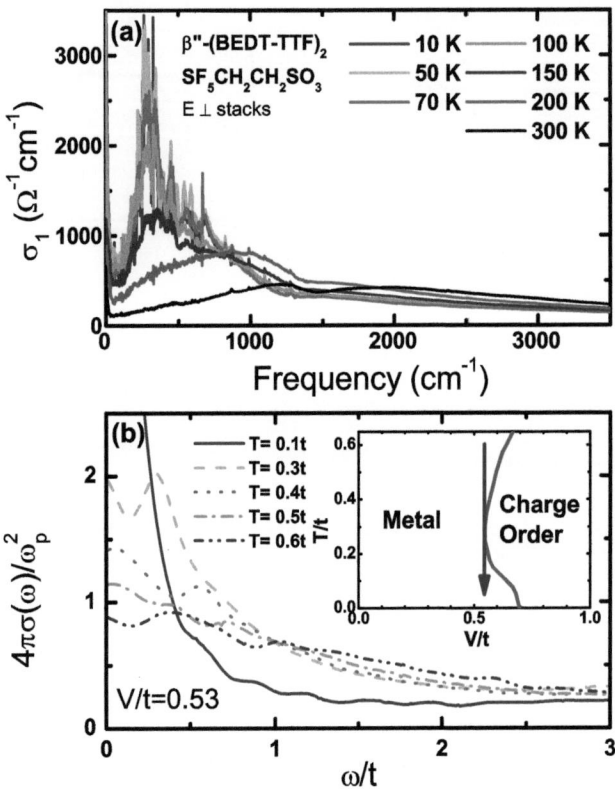

Figure 10.21.: Temperature dependence of the low frequency in-plane conductivity of β''-$(BEDT-TTF)_2SF_5CH_2CF_2SO_3$. Evolution of the charge fluctuation band. (a) Experimental response measured perpendicular to the stacks. (b) Optical conductivity of a quarter-filled electron system calculated by the extended Hubbard model for different temperatures using $(U \rightarrow \infty)$ and $V/t = 0.53$. This corresponds to the line shown in the $V-T$ phase diagram of the inset where also boundary between the metallic and charged-ordered phases is indicated. Calculations by A. Greco.

10.2. Temperature dependence of the in-plane optical response

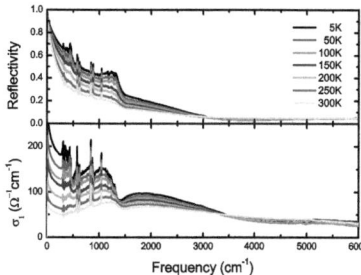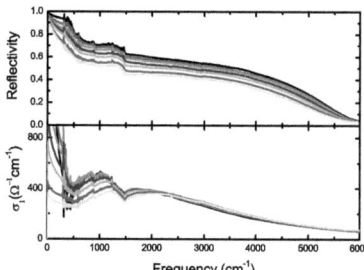

Figure 10.22.: Temperature dependacne of the in plane reflectivity and optical conductivity of the β''-(BEDT-TTF)$_2$SF$_5$CHFSO$_3$ along (a) and perpendicular (b) to the stacks.

As in the experiments (Fig. 10.21 (a)), the low energy feature softens with temperature. For $T/t = 0.1$ the system is located below the back-bending of the diagram where the low energy peak weakens compared to the zero frequency Drude response and the spectra behaves more coherent. The experiment shows this reentrant behavior below $\approx$70-50 K. A value of $t \sim 25$ meV then fits the temperature dependence with the back bending of the charge order transition line. However, using this transfer integral is by a factor of 2-4 lower than expected for β''-(BEDT-TTF)$_2$SF$_5$CH$_2$CF$_2$SO$_3$ but still within the right order of magnitude (Sec. 4.1.1). Then the energy position of the low energy peak is also lower than in the experiment, but considering the simplicity of the model used the agreement is satisfactory. The present theoretical results agree better with optical experiments along the a-direction where, in addition to the lower position caused by the smaller transfer integrals, the low energy features are weaker than along b-direction. Finally, we conclude that the theoretical calculations support qualitatively the interpretation of the low energy optical features at ~ 300 cm^{-1} in terms of the charge fluctuation band. It should be stressed that the calculation is describing well the lowest frequencies, only. Influences to the MIR band caused by finite U, which would alter the self energy, are not taken into account.

10.2.2. β''-(BEDT-TTF)$_2$SF$_5$CHFSO$_3$

For the metallic sister compound β''-(BEDT-TTF)$_2$SF$_5$CHFSO$_3$ the temperature dependent reflectivity is shown in Fig. 10.22 along (a) and perpendicular (b) to the stacks together with their derived optical conductivities. Similar to the superconducting sister compound the reflectivity perpendicular to the stacks is higher than measured along the stacks since also here the overlap to adjacent stacks is higher than the one in the

10. Broad band optical measurements

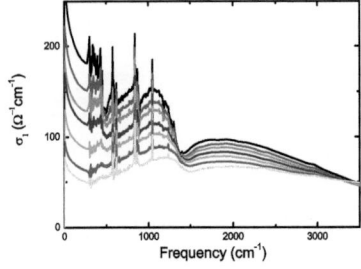
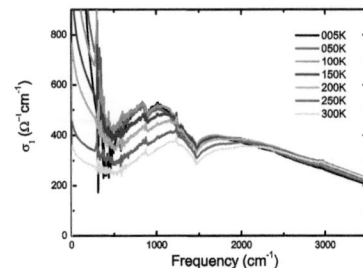

Figure 10.23.: Optical conductivity of the β''-(BEDT-TTF)$_2$SF$_5$CHFSO$_3$ along (a) and perpendicular (b) to the stacks showing the MIR band and onset of the Drude like response.

stack. The increase of reflectivity to lower frequencies as well as the overall increase with decreasing temperature indicate metallic behavior. The reflectivity in the β''-(BEDT-TTF)$_2$SF$_5$CHFSO$_3$ shows a large band located in the MIR range. As for the superconductor perpendicular to the stacks a huge band at 1600 cm^{-1} extends up to 6000 cm^{-1} while a more overdamped behavior is observed along the stacks. Above these band in the NIR-VIS interband transitions are similar as in the superconductor. That is expected since these transitions are between the states in BEDT-TTF molecules which are the building blocks in both compounds and in any other BEDT-TTF charge transfer salts. The transition itself is not influenced by the electronic correlations and so big changes in the interband transitions for different salts are neither expected nor known from literature. As discussed in detail in the following at low frequencies basically an up turn to lower frequencies and increase with decreasing temperature is observed. In the optical conductivity the MIR range is occupied with a large band increasing with decreasing temperature. In comparison to the superconductor it does not become so dominating. On cooling spectral weight shifts to low frequencies. For temperatures below 150 K at frequencies around 350 cm^{-1} the slight onset of a band comparable to the charge fluctuation band in the superconductor can be seen. At lowest frequencies the tendency to a Drude response is observed.

The MIR band

At room temperature the MIR band is located around 1700 cm^{-1} perpendicular and around 1400 cm^{-1} along the stacking direction. As for the superconductor this band can be assigned to the Hubbard like charge transfer band. It shows a red-shift with decreasing temperature (Fig. 10.23). That is the same behavior found for the superconductor and

10.2. Temperature dependence of the in-plane optical response

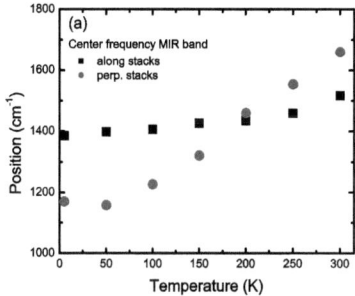
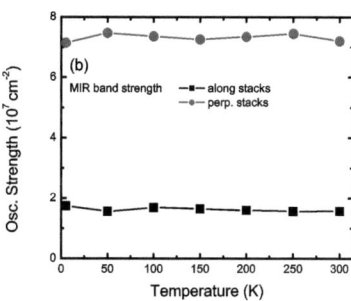

Figure 10.24.: (a) Position and (b) spectral weight (represented by its oscillator strength) of the MIR band in β''-(BEDT-TTF)$_2$SF$_5$CHFSO$_3$. The direction along the stacks is given as red dots and the values perpendicular to the stacks as black squares.

is attributed to the change of the effective correlations caused by the increased overlap between the sites on shrinking of the crystal with decreasing temperature. The shift is about 650 cm^{-1} for the direction perpendicular to the stacks while it is around 240 cm^{-1} along the stacks (Fig. 10.24 (a)). Again, that shows the stronger shrinking of the crystal perpendicular to the stacks. The spectral weight of the bands shows no significant changes in the temperature dependence (Fig. 10.24 (b)). There is no change of the system due to localization effects. The transfer of spectral weight to lower frequencies is mainly due to the red-shift of the MIR band. The spectral weight that should increase due to the shrinking volume of the crystal is fully transfered into the coherent part. That suggests that the main electron dynamics takes place just in the quasi free charge carriers. As seen for the superconducting sister compound, there is a huge influence to the band due to e-mv coupled phonons. They are modeled by a Fano line-shape. The ones with most influence are the modes at 1471 cm^{-1} and 889 cm^{-1}. The other vibrational features are either less coupled to the electronic system or belong to the anion subsystem. They appear as narrow bands superimposed to the electronic spectrum.

The low frequency response

As the measured single crystals were of a small size, the low frequency range accessible is limited. Therefore it is difficult to make a distinct statement about the Drude behavior. The low frequency part of the MIR band perpendicular to the stack direction (Fig. 10.23 (a)) shows an upturn in conductivity below 500 cm^{-1} for all temperatures. The upturn gets more pronounced as the temperature is decreased. Extrapolating this increase with

10. Broad band optical measurements

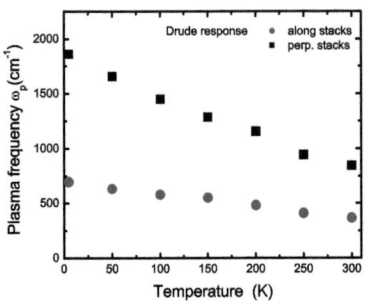

Figure 10.25.: Temperature dependence of the plasma frequency extracted from the partial sum rule over the low frequency spectral weight perpendicular (black square) and along the stacks (red dots) in β''-(BEDT-TTF)$_2$SF$_5$CHFSO$_3$.

the Hagen-Rubens relation to lowest frequencies describes an increase of a metallic like Drude response. Along the stacks the upturn is not that obvious (Fig. 10.23 (b)). Only below 200 K it is noticeable. But the increase of the Drude compound there accounts for the increase of the overall conductivity at low frequencies.

No exact fit to a Drude peak is possible. Therefore the plasma frequency can be extracted only via a partial sum rule from the low spectral weight assumed in the zero frequency peak. To see a tendency, the total extrapolated spectral weight at frequencies below the lowest measured frequency of 288 cm^{-1} was assigned to the Drude contribution. The result is shown in Fig. 10.25. Along both directions it shows an increase of spectral weight at low frequencies which in this case is attributed mainly to the shrinking of the crystal and an increased metallic behavior. No significant spectral weight shift due to correlation effects that lead to localization is seen.

That is in a perfect agreement with the metallic in-plane transport properties. The same holds for the vibrational properties where no traces for localization are seen in IR and only some very slight charge redistribution is observed in Raman. To note is that latest measurements of N. Drichko confirmed the behavior shown here by the extrapolation. All low frequency dynamics shows a Drude response.

The charge fluctuation band

Even though the main increase of the band is attributed to the metallic response there is a slight onset of a charge fluctuation feature. It is visible in both directions below 350 cm^{-1} (Fig. 10.26). It develops at 150 K perpendicular and below 100 K along the stacks and strengthens slightly. But compared to the superconductor the band sits on the

10.2. Temperature dependence of the in-plane optical response

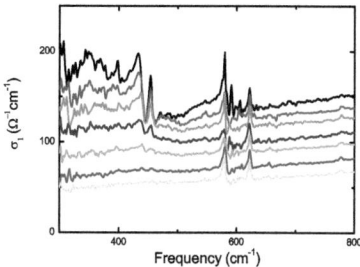
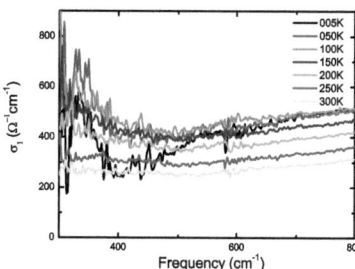

Figure 10.26.: Low frequency optical conductivity of the β''-(BEDT-TTF)$_2$SF$_5$CHFSO$_3$ along (a) and perpendicular (b) to the stacks showing the onset of the charge fluctuation band below 350 cm^{-1} at low temperatures.

background of a Drude response instead within a pseudogap. Also it does not strengthen significant enough that it would dominate the spectra.

10.2.3. Comparison

The room temperature properties (Fig. 10.2) of the two systems, the superconductor β''-(BEDT-TTF)$_2$SF$_5$CH$_2$CF$_2$SO$_3$ and its isostructural metal β''-(BEDT-TTF)$_2$-SF$_5$CHFSO$_3$, are very similar. Both show basically the same electronic band structure. Also the increasing Drude spectral weight on decreasing temperature is present in both systems (Figs. 10.10 and 10.25). Further, some features in the temperature dependent properties which especially can be assigned to structural features show similarities: Both systems shrink on cooling which lead to a redshift of the MIR bands (Figs. 10.8 (a) and 10.24 (a)). Also both shrink more perpendicular to the stacks. However the metal not that significantly along the stacks. Others electronic properties alter significantly due to the different electron-electron correlations in β''-(BEDT-TTF)$_2$SF$_5$CH$_2$CF$_2$SO$_3$ and β''-(BEDT-TTF)$_2$SF$_5$CHFSO$_3$. The MIR Hubbard like charge transfer band which represents the correlation influence is nearly temperature independent in the metallic β''-(BEDT-TTF)$_2$SF$_5$CHFSO$_3$ (Fig. 10.24 (b)). That is in agreement with the nearly perfect metallic character. On the other hand in the superconductor β''-(BEDT-TTF)$_2$-SF$_5$CH$_2$CF$_2$SO$_3$ the metallic properties are also present. They show up in a strongly renormalized Drude response but at the same time the system shows traces of charge order. Its charge transfer band shows a significant temperature dependence with increasing spectral weight on cooling (Fig. 10.8 (b)). Below 125-150 K charge fluctuations set in. Right at that temperatures a drop in the MIR spectral weight takes place and is transfered into a low frequency feature (Fig. 10.20) which is associated to charge fluctu-

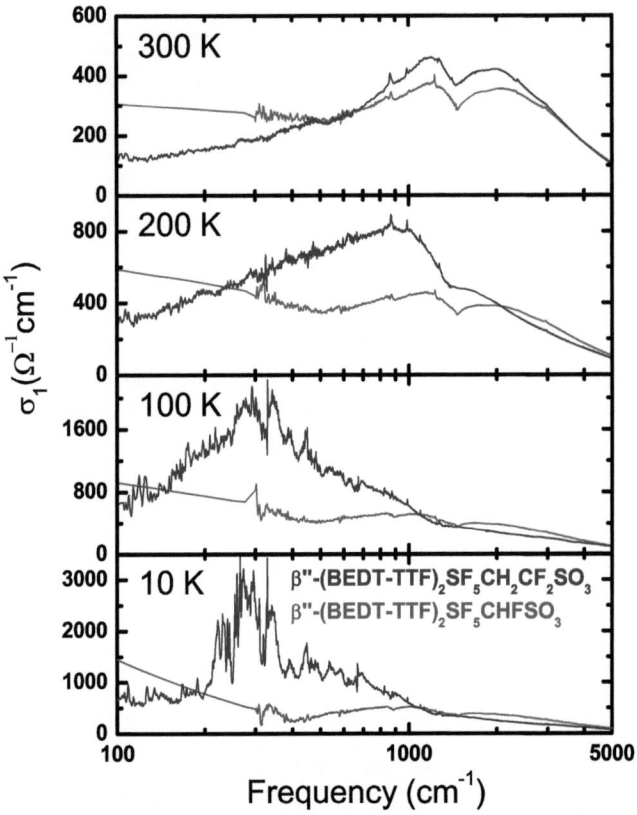

Figure 10.27.: Comparison of the optical conductivity along the interstack direction (b-axis) in β''-(BEDT-TTF)$_2$SF$_5$CH$_2$CF$_2$SO$_3$ (blue) and β''-(BEDT-TTF)$_2$-SF$_5$CHFSO$_3$ (red) at different temperatures.

ations. In addition a low frequency lattice phonon is present which can be interpreted as a collective charge-order excitation (Fig. 10.16).

The in-plane conductivity for both compounds is compared in Fig. 10.27. At room temperature in the superconductor β''-(BEDT-TTF)$_2$SF$_5$CH$_2$CF$_2$SO$_3$ the main part of the spectrum exhibits a localized character. Its Drude response starts well below the shown frequency range. However in the metallic sister compound β''-(BEDT-TTF)$_2$-SF$_5$CHFSO$_3$ already below 500 cm^{-1} the upturn to a Drude response is obvious. On cooling the low frequency spectral weight enhances for the superconductor and the charge fluctuation band around 300 cm^{-1} appears. In the metal mainly the Drude response is enhanced. The lack of the strong charge fluctuation band in the less correlated β''-(BEDT-TTF)$_2$SF$_5$CHFSO$_3$ metal evidences the absence of strong interactions between the charge carriers through charge fluctuations. That is in agreement with the vibrational analysis where no traces of fluctuations are found. Due to that no attractive interaction is mediated and no superconductivity is supported; the system stays metallic.

10.3. The superconductor β-(EDT-TTF)$_4$[Hg$_3$I$_8$]$_{(1-x)}$

The optical properties of the superconductor β-(EDT-TTF)$_4$[Hg$_3$I$_8$]$_{(1-x)}$ are measured along all principal axes. It should be noted that only 2 samples could be measured and further that the sample used for the in-plane properties along a- and b-axis is different from the one measured in the vibrational dependence along the insulating c-direction. After the measurements, the samples were characterized by DC resistivity; for details see Sec. 8.4. The sample used in the vibrational analysis stayed metallic and turned superconducting at about 8 K. The sample measured for the in-plane properties turned insulating at low temperatures.

For room temperature, the reflectivity and calculated conductivity are given in Fig. 10.28. In principle the electronic properties are similar to the BEDT-TTF salts since the orbital overlap of the EDT molecules also leads to a quasi-2D system. The main differences are due to the lower molecular symmetry and therefore are mainly in the molecular vibrations. In the electronic picture the different molecular orbital of EDT-TTF compared to BEDT-TTF will change especially the on-site Coulomb repulsion.

In detail, the room temperature behavior of the β-(EDT-TTF)$_4$[Hg$_3$I$_8$]$_{(1-x)}$ is indeed similar to the β''-(BEDT-TTF)$_2$SF$_5$CH$_2$CF$_2$SO$_3$ in its electronic band structure, which is dominated by a strong MIR band. One can clearly see the two dimensional character of the β-(EDT-TTF)$_4$[Hg$_3$I$_8$]$_{(1-x)}$. The reflectivities in the ab-plane are significantly higher than the non conducting c-direction. The in-plane directions show band properties discussed in the following while the insulating c-direction shows no presence of electronic bands. Vibrational features along this direction were discussed in Sec. 9. The in-plane out-of-plane anisotropy $\sigma_{ab}/\sigma_c \approx 100$ for the β-(EDT-TTF)$_4$[Hg$_3$I$_8$]$_{(1-x)}$ is high like in the BEDT compounds. The in-plane anisotropy of $\sigma_a/\sigma_b \approx 3$ is comparably small. The MIR band in the high conducting b-direction is located at 1500-2000 cm^{-1} with an roll of in reflectivity at around 5000 cm^{-1}. Also in conductivity the band is well

10. Broad band optical measurements

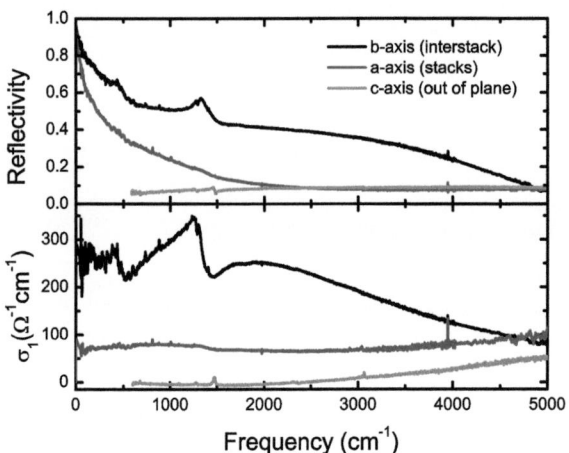

Figure 10.28.: Reflectivity and conductivity of the β-(EDT-TTF)$_4$[Hg$_3$I$_8$]$_{(1-x)}$ organic superconductor along the principal axes at room temperature.

10.3. The superconductor β-(EDT-TTF)$_4$[Hg$_3$I$_8$]$_{(1-x)}$

developed. Along the a-direction, perpendicular to the stacks, the MIR band is not so pronounced anymore. Only a very flat band shows up in conductivity. Its over-damped character is much stronger than observed in the BEDT-TTF salts as it is very well seen in the reflectivity spectra. Due to the molecular symmetry no strong influence of the e-mv coupled A_g equivalent of the BEDT vibrational mode is seen around 1400 cm^{-1} perpendicular the stacks. For the BEDT-TTF salts this coupling causes the strong Fano line-shape. In EDT-TTF along the stacks the feature at 1450 cm^{-1} is still present. The overall frequency dependence is metallic, the increase of reflectivity to lower frequencies is observed along both directions. However a distinct Drude peak in conductivity at lowest frequencies cannot be seen.

The MIR band

The in-plane temperature dependence of the reflectivity and conductivity of the β-(EDT-TTF)$_4$[Hg$_3$I$_8$]$_{(1-x)}$ organic superconductor is given in Fig. 10.29. An overall increase of reflectivity with decreasing temperature is seen for both directions. In conductivity this corresponds to an increase of the broad MIR band around 1400 cm^{-1} along the stacks. Its position and strength is given in Fig. 10.30. A hardening till 150 K and a softening on further cooling is observed. The bands spectral weight increases linear till 150 K, then the increase flattens. On cooling below 12 K a strong increase is observed. That is right the temperature range when the sample turned insulating in the DC resistivity. However, in the interstack direction no distinct MIR is observed. Here around 2000 cm^{-1} a minimum in the conductivity is observed. Only the low frequency response can be successfully analyzed. A band is present around 200 cm^{-1} and increases on cooling. It is also observed in the stack direction. It is located in the frequency range of the charge fluctuation band in the BEDT-TTF based compounds.

The low frequency response

The conductivity along and perpendicular to the stacks for low frequencies is plotted in Fig. 10.31. It shows the evolution of the low frequency band around 200 cm^{-1}. It is present already at room temperature. Its strength is given in Fig. 10.32. Along the stacks (black squares) the band first slightly increases on cooling. When the system turns insulating below 12 K the band significantly increases. In the interstack direction (red circles) the band increases on cooling till 100 K. Then it decreases and enhances again below 10 K. Comparing both directions at 3 and 6 K one could speculate about the evolution of a phonon like feature around 60 cm^{-1}. It enhances for both directions. Assigning the feature to a simple Drude increase at low frequencies would be in contradiction to the DC data. It can be fitted as a peak at 100 cm^{-1} along and at around 52 cm^{-1} in the interstack direction (Fig. 10.32 (a)). In the stack direction it carries significant spectral weight as seen by its oscillator strength (Fig. 10.32 (b)) represented as blue triangles. In interstack direction the peak carries only a small spectral weight shown as green diamonds.

At lowest frequencies no distinct Drude response is found and no integrated spectral

10. Broad band optical measurements

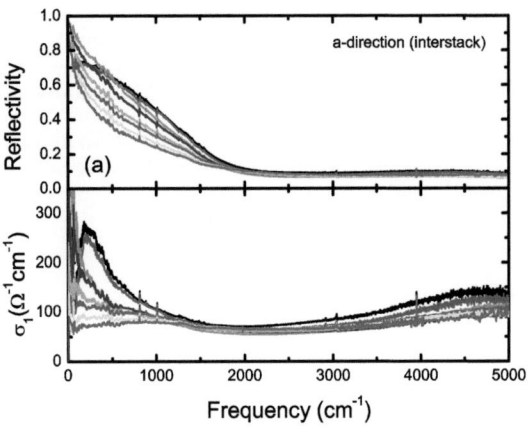

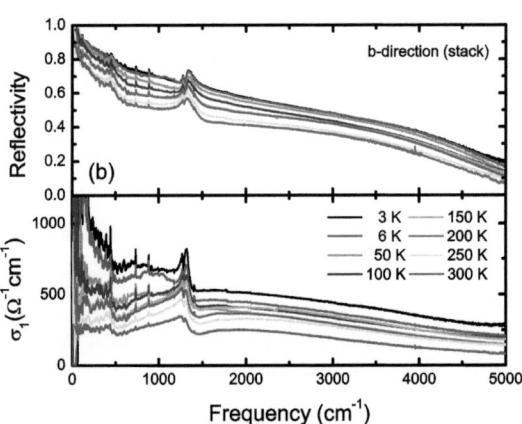

Figure 10.29.: Temperature dependence of the in plane reflectivity and optical conductivity of the β-(EDT-TTF)$_4$[Hg$_3$I$_8$]$_{(1-x)}$ (a) perpendicular to and (b) along the stacks.

10.3. The superconductor β-$(EDT$-$TTF)_4[Hg_3I_8]_{(1-x)}$

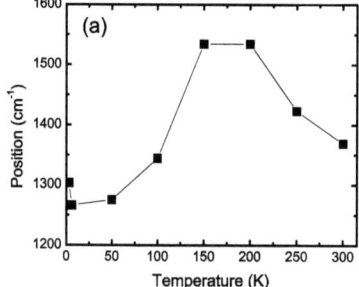 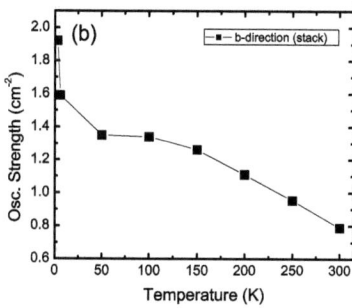

Figure 10.30.: (a) Position and (b) spectral weight (represented by its oscillator strength) of the MIR band in β-$(EDT$-$TTF)_4[Hg_3I_8]_{(1-x)}$ along the stacks.

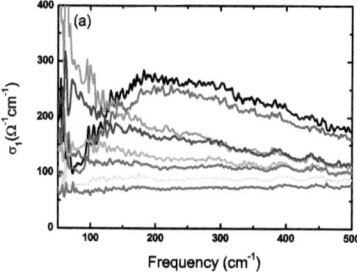 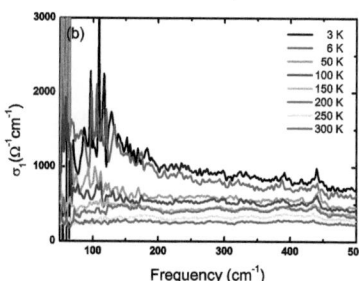

Figure 10.31.: Temperature dependence of the low frequency in-plane optical conductivity of the β-$(EDT$-$TTF)_4[Hg_3I_8]_{(1-x)}$ (a) perpendicular to and (b) along the stacks.

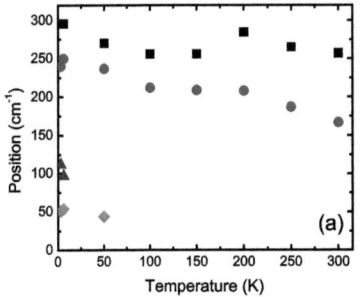 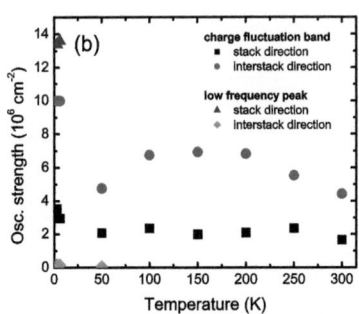

Figure 10.32.: (a) Position and (b) strengths of the charge fluctuation band in β-(EDT-TTF)$_4$[Hg$_3$I$_8$]$_{(1-x)}$ along the stacks (black squares) and in the interstack direction (red circles). Below 50 K an additional low frequency peak appears.

weight can be extracted. However based on the metallic Hagen-Rubens extrapolation Eqn. 2.3 and the Drude-Lorentz fit (Sec. 2.1.1) the plasma frequency is obtained (Fig. 10.33). Along both directions the plasma frequency increases on cooling. Along the stacks (black squares) there is a strong enhancement below 200 K. In the interstack direction it is a more gradual increase down to 100 K and stronger increase to 50 K. At 3 and 6 K no metal like response is found anymore. The corresponding spectral weight is transfered to the low frequency peak, the charge fluctuation, and also the MIR band. That behavior is in perfect agreement with the transport properties measured on that sample. As described in Sec. 8.4 this sample became insulating below 12 K. The reason for that is unclear. One possibility is that the system is very close to the charge order transition. Smallest changes can push it to the transition. Also the micro-cracks, seen in the jumps in the DC properties, could lead to defects in the crystal that cause stress or strain. Some other samples of that batch even cracked completely into pieces on cooling. Like known from the very sensitive pressure dependence of the system [134, 136] such micro-cracks could lead into a condition of a metal-insulator transition at 12 K. However, a final statement cannot be made.

10.4. THz spectroscopy of the superconducting gap

THz spectroscopy is used for two reasons. First as low frequency extend to the reflectivity data which enables to see the low frequency Drude response in β''-(BEDT-TTF)$_2$SF$_5$CH$_2$CF$_2$SO$_3$. The main feature in the THz range is the superconducting gap. Two different THz schemes are used. Absolute reflectivity measurements (Sec. 6.1.2)

10.4. THz spectroscopy of the superconducting gap

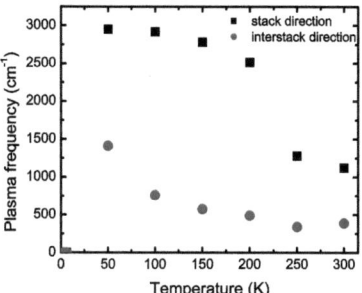

Figure 10.33.: Temperature dependence of the plasma frequency extracted from the fit of the low frequency response in β-(EDT-TTF)$_4$[Hg$_3$I$_8$]$_{(1-x)}$. Below 12 K no Drude response is found. The sample turns insulating.

show the superconducting gap in the conductivity spectra. Relative reflectivity measurements trace the development of the gap depending on temperature and magnetic field (Sec. 6.1.2).

10.4.1. Absolute reflectivities in the superconducting gap regime

The absolute reflectivity for β''-(BEDT-TTF)$_2$SF$_5$CH$_2$CF$_2$SO$_3$ are measured above (10 K) and well below (1.75 K) the superconducting transition temperature ($T_c \approx 5.4$ K). The results are presented in Fig. 10.34. Above the superconducting gap frequency the reflectivity is the same for both temperatures, right above and below the superconducting transition. In the superconducting state the reflectivity goes up to 100% for frequencies below the superconducting gap. For β''-(BEDT-TTF)$_2$SF$_5$CH$_2$CF$_2$SO$_3$ the gap frequency can be estimated to be $2\Delta \approx 12$ cm^{-1} from Fig. 10.34. Below the gap the reflectivity in the superconducting state is flat and is approximatly equal to 1. Right at the gap at 12 cm^{-1} there is a feature which might be a phonon. However the noise level of the measurement is very high and the assumed level of $R = 1$ even above 1. The reflectivity ratio between the temperatures is calculated in Fig. 10.35. The reflectivity enhancement below the gap is around 10-12 cm^{-1} shown by the red and green line drawn as guide to the eye. From that the reflectivity in the normal state is approximately 4% below reflectivity of the superconducting state. To estimate the gap in conductivity, the optical conductivities at in the normal conducting (10 K) and superconducting state (1.7 K) are compared. Since the 1.75 K was measured only in the low frequency region the 10 K data was used as 'mid- and high frequency extrapolation' to that. In the low frequency regime the extrapolation is done with Hagen-Rubens in the metallic regime

10. Broad band optical measurements

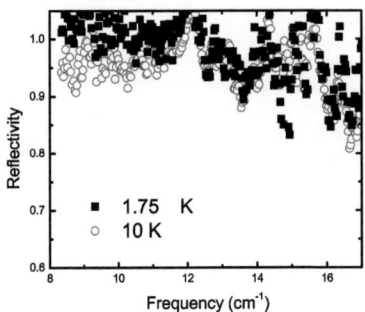

Figure 10.34.: Absolute reflectivity of the β''-(BEDT-TTF)$_2$SF$_5$CH$_2$CF$_2$SO$_3$ above and below the superconducting transition at T$_c$=5 K.

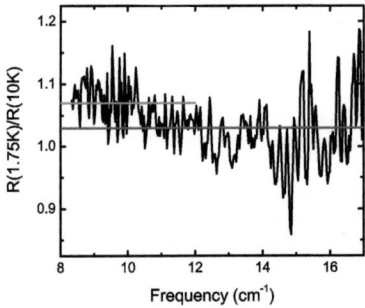

Figure 10.35.: Reflectivity ratio R(1.75K)/R(10)K of β''-(BEDT-TTF)$_2$SF$_5$CH$_2$CF$_2$SO$_3$. An enhancement of the reflectivity by approximately 0.04 (from the red to the green line drawn as guide to the eye) is found between 10-12 cm^{-1}.

10.4. THz spectroscopy of the superconducting gap

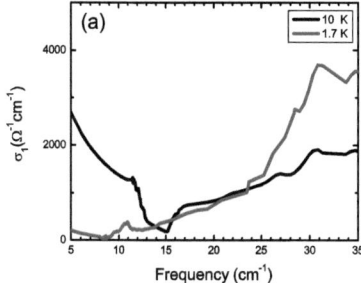
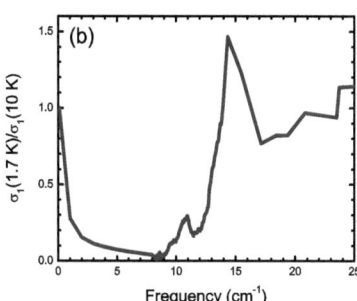

Figure 10.36.: The superconducting gap in β''-(BEDT-TTF)$_2$SF$_5$CH$_2$CF$_2$SO$_3$. (a) The optical conductivity at 10 and 1.75 K. (b) Conductivity ratio between superconducting and normal conducting state. Due to the high noise level at the low frequencies as seen in Figs. 10.34 and 10.35 the data here is highly smoothed with a 70 point adjacent average filter.

and with R=1 for the superconducting reflectivity. Fig. 10.36 (a) shows both conductivities. Taking the ratio of them results in the graph (b). One can clearly see the full superconducting gap opening below 12 cm^{-1}, the superconducting conductivity drops compared to the normal state conductivity. The spectral weight is shifted to a delta peak at zero frequency. The kink feature at 15 cm^{-1} is due to the high noise in the low frequency data and not considered to be real. Above the gap the conductivities are basically the same, the ratio is nearly 1. However, above 25 cm^{-1} the ratios are not the same anymore since the strong influence of the low frequency lattice phonon comes into play.

The absolute measurements of the reflectivity presented here show that the behavior of the reflectivity is in agreement with the expected behavior of a superconductor. The conductivity data shows the superconducting gap at low frequencies. Observing the full gap open in optics shows that there are no nodes present and s-wave symmetry is suggested. To further prove that these effects are related to superconductivity the temperature dependence of the optical gap and the influence of magnetic field is measured by relative reflectivity changes.

10.4.2. Opening of the superconducting gap

To investigate the opening of the superconducting gap, the reflectivities with respect to the normal state reflectivity right above the transition temperature are measured. Fig. 10.37 (a) shows the reflectivity ratios in the superconducting gap regime. Measure-

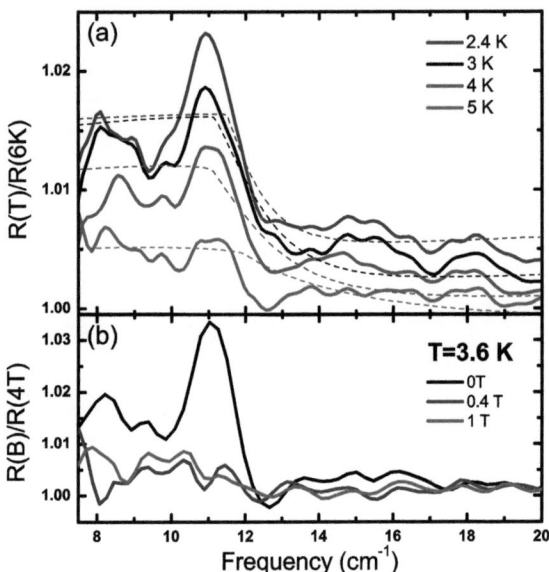

Figure 10.37.: (a) Temperature dependent opening of the superconducting gap. Reflectivity ratios are taken with respect to the 6 K reflectivity which represents the normal state. The different lines are offset to each other by 0.002 for clarity. (b) Closing of the gap by an applied magnetic field at T=3.6 K. Ratios are taken versus the reflectivity at 4T representing the normal state. All data is smoothed with a 20 point adjacent average filter.

10.4. THz spectroscopy of the superconducting gap

ments were taken at 15, 10, 6, 5, 4, 3, and 2.4 K. Referencing the reflectivities to 15 K showed just flat lines for the $R(10\ \text{K})/R(15\ \text{K})$ and $R(6\ \text{K})/R(15\ \text{K})$ ratios (not shown). Therefore all ratios in the investigation of the superconducting gap are taken against the 6 K reflectivity as normal state. Below the superconducting transition temperature ($T_c \approx 5.4$ K) the reflectivity increases below the gap frequency of $2\Delta \approx 12$ cm^{-1} and stays equal to the normal state reflectivity for frequencies above the gap. The reflectivity in the gap clearly enhances with decreasing temperature. This shows the opening of the gap. The enhancement reaches from ≈ 0.005 at 5 K to ≈ 0.015 at 2.4 K. From the absolute reflectivity measurements it can be estimated to 0.04 at 1.8 K as described above (Fig. 10.35). Since the superconducting gap frequency and the scattering rate are of the same order of magnitude the system is neither in the clean nor in the dirty limit. Therefore a simple Mattis-Bardeen description [90] is not possible. That is already seen in the conductivity (Fig. 10.36) of the normal state which has still a strong frequency dependence within the range of the superconducting gap. In Fig. 10.37 the dashed lines are fits which were performed based on BCS theory in a formulation valid for arbitrary purities [91]. The resulting gap frequency delivers $2\Delta \approx 12$ cm^{-1} in agreement with the absolute reflectivity measurements. A temperature dependence of the gap value cannot be extracted due to a strong error of estimating the onset of the enhancement. Right there, as already seen in the absolute reflectivity values, a phonon like feature is located. The purity/gap ratio extracted from the fits is about $2\Delta/\tau^{-1} \sim 1$ and the gap/temperature ratio results in $2\Delta/(k_b T_c) = 3.3$. That is close to the ratio of 3.5 expected for a BCS-type superconductor.

To prove that the enhancement of reflectivity is definitely related to the superconductivity, an magnetic field is applied to break the superconducting state. For the magnetic field dependence, shown in Fig 10.37 (b), all spectra are taken at the same temperature (T=3.6 K) for 0, 0.4, 1, and 4T. The critical field is 3.4 T [128]. The ratios are referenced with respect to the 4 T measurements which represents the normal state. For zero field vs. 4T the reflectivity enhancement is about 0.015. That is in agreement with the temperature dependent measurements. At fields of 0.4 and 1 T the gap is nearly closed, the enhancement is ≈ 0.005 at maximum, only. That shows that the gap is nearly closed already for small fields.

10.4.3. Comparison to the superconducting gap in α_t-(BEDT-TTF)$_2$I$_3$

The optical properties of the gap observed here can be compared to the superconducting gap observed in the tempered α_t-(BEDT-TTF)$_2$I$_3$ [185]. It becomes superconducting at around T_c=8 K. Its optical properties are presented in Fig 10.38. On the left side the absolute values of reflectivity (a) and optical conductivity (b) for 15 K (normal state) and 3.6 K (superconducting state) are shown. On the right side the reflectivity (c) and conductivity ratios (d) of the superconducting state versus the normal state are presented. The optical gap is around 25 cm^{-1}, extracted from a BCS fit to the conductivity ratio. $2\Delta/k_b T_c$=4.4 puts the system into the moderate coupling regime. In the gap the absolute reflectivity in the superconducting state enhances by about 0.05 compared to

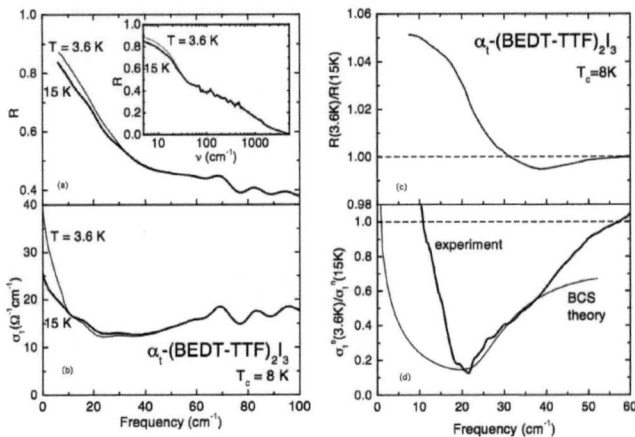

Figure 10.38.: The BCS like superconducting gap in α_t-(BEDT-TTF)$_2$I$_3$ (T_c=8 K). Absolute (a) reflectivity (the inset shows the and (b) conductivity of the normal (15 K) and superconducting state (3.6 K). The relative ratios of the superconducting reflectivity and conductivity normalized by the normal state properties are given in (c) and (d) respectively. A superconducting gap frequency of 25 cm^{-1} is extracted from a BCS fit to the conductivity ratio. From [185].

the normal state reflectivity. Slightly above the gap, between 30-60 cm^{-1} the reflectivity of the superconducting state is even a bit lower than the normal state reflectivity. The same behavior is observed in β''-(BEDT-TTF)$_2$SF$_5$CH$_2$CF$_2$SO$_3$ with the gap at 12 cm^{-1}. Within the gap the reflectivity enhances by about 0.04 and also above the gap in the region from 12-16 cm^{-1} the superconducting reflectivity is be slightly below the normal state reflectivity. The temperature dependence of the gap was not investigated for α_t-(BEDT-TTF)$_2$I$_3$. The opening of the optical gap as traced here for β''-(BEDT-TTF)$_2$SF$_5$CH$_2$CF$_2$SO$_3$ was not measured on an organic superconductor so far. Also the closing of the optical gap with an applied magnetic field was not observed before.

In conductivity the optical gap is clearly seen, too. For the α_t-(BEDT-TTF)$_2$I$_3$ the superconducting conductivity starts to drop below the normal state conductivity below 55 cm^{-1}. The lowest superconducting conductivity is found around 20 cm^{-1} with about 0.2 of the normal state conductivity before it recovers again. The not completely open gap is attributed to the fact that only a fraction of the system became superconducting while a part of it still stayed normal conducting. In the β''-(BEDT-TTF)$_2$SF$_5$CH$_2$CF$_2$SO$_3$ the gap properties show a similar behavior. Below 25 cm^{-1} the conductivity of the superconducting state drops below the normal state conductivity with the lowest conductivity around 8 cm^{-1}. Here the conductivity essentially drops to zero so the gap is completely open. However, one has to keep in mind that this frequency corresponds to the low-frequency limit which was accessed with the method of measuring absolute reflectivity values. Frequencies below are extrapolated.

In total, both organic superconductors show an optical gap in the superconducting state that can be described by a BCS model. The higher T$_c$ in the α_t-(BEDT-TTF)$_2$I$_3$ is likely due to the fact that it is closer to the charge order transition. The non-tempered system α-(BEDT-TTF)$_2$I$_3$ is a perfect charge ordered insulator at low temperatures. However, β''-(BEDT-TTF)$_2$SF$_5$CH$_2$CF$_2$SO$_3$ shows a completely open gap and is closer to a typical BCS behavior.

Part IV.

Discussion

11. The Interplay of Charge Order and Superconductivity

When we summarize the findings of the in-plane optics in Sec. 10 and compare it with the results of the vibrational spectroscopy in Sec. 9 and the characterization of the transport properties in Sec. 8, it becomes evident that there is an interplay of charge order and superconductivity. The significant features are in a good agreement with a theoretical picture of correlation driven charge fluctuations (Sec. 3) which also describes the possibilities of superconductivity. The comparison of the characteristic features for different correlated electron compounds evidences their connection to the ground state of the system. Additional findings and deviations from the proposed model help to understand its limitations. That paves the way to understand the features in a more general framework of correlated electron systems.

11.1. The charge ordered metallic state and superconductivity

The picture of charge fluctuations in a charge-ordered metallic state applies for the quarter-filled two-dimensional organic superconductors. The β''-(BEDT-TTF)$_2$SF$_5$CH$_2$CF$_2$SO$_3$ investigated as prime example. The comparison to the metals, in particular β''-(BEDT-TTF)$_2$SF$_5$CHFSO$_3$, points out their influence on superconductivity. First of all, the investigated superconductor and the isostructural metal show no structural changes. Therefore all observed features are due to electronic properties. The possibility of the enhancement of electronic effects due to the lattice dynamic is discussed in a subsequent section. Here the characteristic features observed in the experiments are compared to related materials but also to completely different classes of correlated systems, like manganites or cuprates.

11.1.1. Transport properties and the low-frequency response of bad metals

The organic superconductor β''-(BEDT-TTF)$_2$SF$_5$CH$_2$CF$_2$SO$_3$ has metallic DC properties down to the superconducting transition but in addition shows the presence of charge order fluctuations for temperatures below ≈ 150 K. The microwave transport properties give a first hint to charge fluctuations. They show deviations from a simple metallic response and lead to an upturn in the microwave-resistivity before the system enters the superconducting state. Similar resistivity upturns are known for several organic

11. The Interplay of Charge Order and Superconductivity

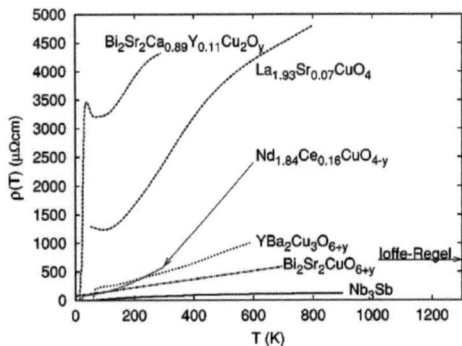

Figure 11.1.: Resistivity of several copper oxides. $La_{1.93}Sr_{0.07}CuO_4$ (LSCO) and $Bi_2Sr_2Ca_{1-x}Y_xCuO_2$ show a resistivity upturn at low temperatures. At high temperatures they are discussed because no resistivity saturation at the Ioffe-Regel limit (marked by the arrow for LSCO) is observed. From [56].

conductors in DC properties (see table and references in [17]). This behavior in DC-resistivity is also observed in the bad metal phase of low doped $La_{2-x}Sr_xCuO_4$ (LSCO) [186] like $La_{1.94}Sr_{0.06}CuO_4$ [187], and $La_{1.93}Sr_{0.07}CuO_4$ [188], or in other cuprates like $Bi_2Sr_2Ca_{1-x}Y_xCuO_2$ [102]. The latter two are shown in Fig. 11.1 together with other systems. They are discussed in the framework of resistivity saturation beyond the Ioffe-Regel limit at high temperatures under the influence of high correlations [56]. There the suppression of the Drude peak in favor of incoherent low frequency contributions is suggested to be due to electron-electron and/or electron-phonon interactions by reducing the magnitude of the hopping energy. That might be relevant for metal-insulator transitions or superconductivity in the low temperature regime as assumed and discussed for alkali-doped fullerenes [189, 190].

Also the organic superconductor β''-$(BEDT-TTF)_2SF_5CH_2CF_2SO_3$ does not saturate in DC-resistiviy up to room temperature and shows a mw-resistivity up-turn right before it becomes superconducting. In addition, characteristic features in its optical response are found which are characteristic for a bad metal [56]: There is only a small Drude peak but large incoherent contributions. These are the charge fluctuation band and a collective mode coupled to a phonon excitation (see Fig. 11.2). The comparison to the isostructural metal β''-$(BEDT-TTF)_2SF_5CHFSO_3$ (Sec. 10.2.3) shows that strong incoherent contributions on expense of the Drude spectral weight are due to higher electronic correlations. The origin of the features proposed here is electron-electron interaction. However, the coupling of the vibrational modes to the lattice via emv-coupling and phonon-modes are relevant, too. Especially the collective mode, which is suggested as a lattice mode coupled to the charge fluctuations, shows the importance of electron-phonon interactions.

11.1. The charge ordered metallic state and superconductivity

In the following the features of the charge-ordered metallic state in the superconductor are compared to the features of the bad-metal state in other correlated systems. Further, their connection to superconductivity is discussed.

11.1.2. The in-plane optical response of the bad metal state

The electronic properties of the charge-ordered metallic phase in the β''-(BEDT-TTF)$_2$-SF$_5$CH$_2$CF$_2$SO$_3$ superconductor are evident in the in-plane optical response. That is shown in Fig. 11.2 and is proposed as a typical response of a strongly correlated quarter-filled system in a charge-ordered metallic state close to the metal-insulator transition. A strong charge transfer band located around $3V$ with an optical pseudo-gap below shows that the electronic system is mainly localized due to the high correlations in the system. A narrow Drude response evidences that a small fraction of charge carriers (for the β''-(BEDT-TTF)$_2$SF$_5$CH$_2$CF$_2$SO$_3$ approximately 5% at 10 K) is participating in the metallic transport, only. At temperatures above 125 K even less.

This is similar to the coherent-incoherent crossover observed for instance in the bad metallic behavior of θ-(BEDT-TTF)$_2$I$_3$ [191] in Fig. 11.3 or for La$_{2-x}$Sr$_x$CuO$_4$ [187] in Fig. 11.4. In both systems the Drude response is present at lowest temperatures, only. On increased temperature the Drude response is suppressed and a finite frequency peak in the FIR appears. The characteristics of this peak are similar to the charge fluctuation band in this thesis. It increases and softens with decreasing temperature. For the θ-(BEDT-TTF)$_2$I$_3$ it appears around 160-320 cm^{-1} and in the case of La$_{2-x}$Sr$_x$CuO$_4$ as broad band at higher frequencies around 800 cm^{-1} which covers basically the whole low frequency regime.

For the β''-(BEDT-TTF)$_2$SF$_5$CH$_2$CF$_2$SO$_3$ superconductor it is located in the frequency range of 150-320 cm^{-1}. However, while in the compounds addressed above, at high temperatures the low frequency band is present instead of a Drude peak, in the superconductor both are present: The charge fluctuation band and a small Drude response. As in the bad metals, also the organic superconductor shows a reentrant behavior to a more metallic state at lowest temperatures. The low frequency band in the bad metals shows the strong incoherent character of the charge carriers due to the strong correlations in the system [56, 187]. Its origin is speculated either purely electronic or also coupled to hidden excitations [191]. In the charge fluctuation band bosonic charge fluctuations could act as such hidden excitations. Then the peak strength can be directly associated with the strength of the quasi-particle interactions which are mediated by the fluctuations. These interactions could lead to superconductivity as described in the theoretical model (Sec. 3 and [18]). Comparing the superconductor β''-(BEDT-TTF)$_2$SF$_5$CH$_2$CF$_2$SO$_3$ to its isostructural metal β''-(BEDT-TTF)$_2$SF$_5$CHFSO$_3$, in the latter no traces of strong charge fluctuations are found. Sec. 10.2.3 compared the behavior of the low frequency features in both systems. It confirms the strong enhancement of the charge fluctuation band in the superconductor at low temperatures while in the metallic compound most spectral weight is shifted to the Drude response.

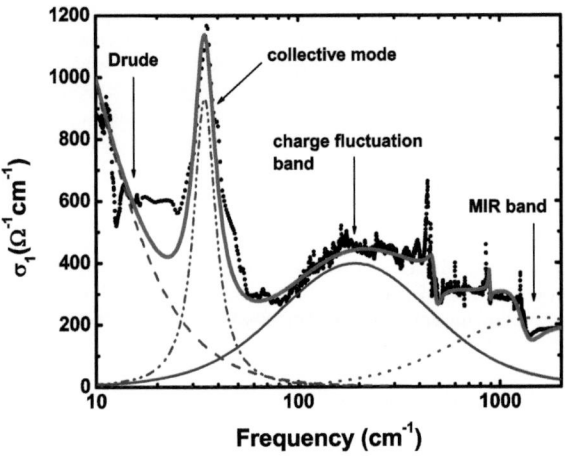

Figure 11.2.: Typical optical response in a quarter-filled two-dimensional organic superconductor at the example of β''-(BEDT-TTF)$_2$SF$_5$CH$_2$CF$_2$SO$_3$. The Drude and the MIR response is shown together with the interaction features arising due to the charge fluctuations in the charge ordered metallic state: A collective excitation of the charge order at 40 cm^{-1} and a charge fluctuation band around 300 cm^{-1}. The measured data is given as black circles described with a Drude Lorentz fit (red line). The single features are separately presented as blue lines. The contributions due to Fano line shaped phonons used in the fit are not shown. Note the logarithmic frequency scale of the plot.

11.1. The charge ordered metallic state and superconductivity

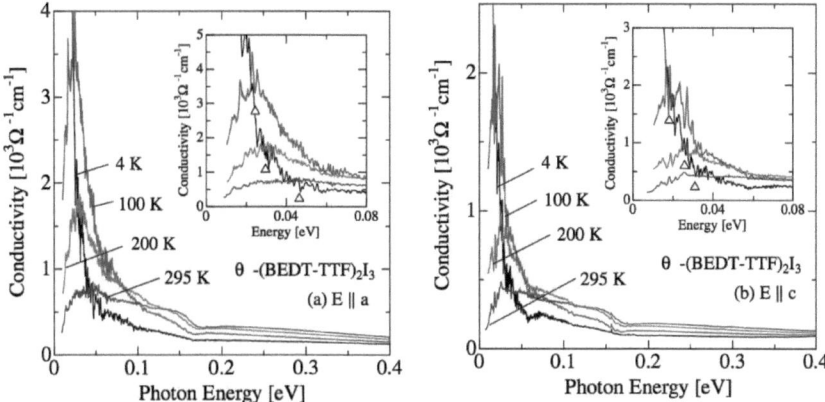

Figure 11.3.: Optical conductivity of θ-(BEDT-TTF)$_2$I$_3$ (a) along a- and (b) along c-direction. At low temperatures a FIR-peak appears and redshifts as indicated by the triangles showing its peak position. The inset shows the magnified low frequency range. From [191].

That these features are related to the interactions of the charge carriers with the fluctuations is also supported by the extended Drude analysis of the coherent response. For the superconductor β''-(BEDT-TTF)$_2$SF$_5$CH$_2$CF$_2$SO$_3$ the scattering rate of the electronic system in the frequency range of the collective mode shows a linear dependence. That reveals an interaction with bosonic excitations, in this case the charge fluctuations. Also for La$_{2-x}$Sr$_x$CuO$_4$ a linear dependence of the scattering rate at low temperatures (Fig. 11.4) suggests a coupling to an collective electronic mode [187]. The corresponding effective mass increases, dragging the cloud of dressing electrons, due to the strong electronic interactions.

In the case of the β''-(BEDT-TTF)$_2$SF$_5$CH$_2$CF$_2$SO$_3$ superconductor the low frequency response in the charge ordered metal regime is similar to the bad-metal regime of other highly correlated systems. The Drude peak is narrow and of small spectral weight, showing that only a small fraction of carriers contributes to the transport. A large amount of spectral weight is transferred to the charge fluctuation band. However, the fact that both features, the Drude spectral weight as well as the interaction features, increase on decreasing temperature shows an interplay of the metallic and charge ordering properties rather than a competition as in the other systems discussed above. There the Drude peak disappears in favor to the incoherent excitation band with increasing temperature. The influence of superconductivity is seen in the optical properties of the coherent carriers. The low frequency response becomes gapped at the transition into the

11. The Interplay of Charge Order and Superconductivity

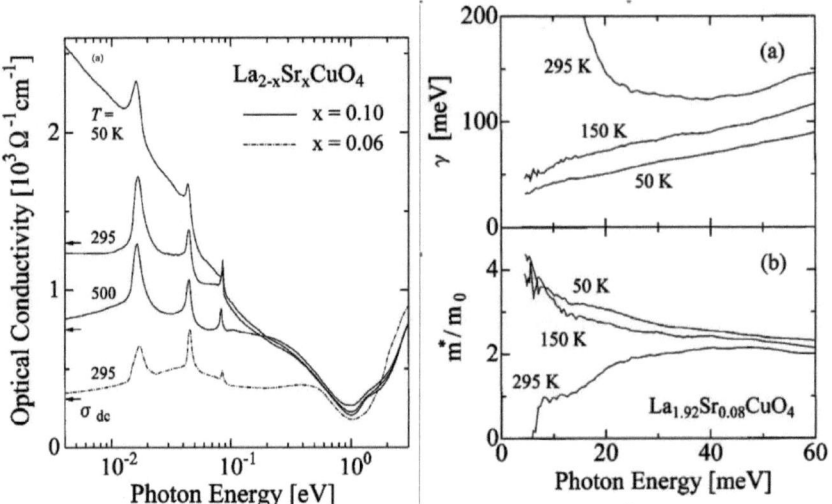

Figure 11.4.: Left: Optical conductivity of $La_{2-x}Sr_xCuO_4$. At low temperatures a Drude peak is present while it disappears at higher temperatures a FIR-band appears. The narrow peaks are phonon features. Right: The frequency dependent scattering rate (a) and the corresponding effective mass (b) for $La_{1.92}Sr_{0.08}CuO_4$. At low temperatures a linear frequency dependence of the scattering rate is found. From [187].

superconducting state.

11.2. The charge fluctuation interaction for different compounds

The main features of the interplay, the charge fluctuation band and the small Drude response, can be identified and compared in the spectra of different organic conductors and superconductors. Again, for the superconductors the close connection to the charge ordered state becomes obvious. In both superconductors, β''-(BEDT-TTF)$_2$-SF$_5$CH$_2$CF$_2$SO$_3$ and β-(EDT-TTF)$_4$[Hg$_3$I$_8$]$_{(1-x)}$, a charge disproportionation is present while the transport properties are metallic. The charge carrier interactions within this state are measured by the strength of the charge fluctuation band.
The β-(EDT-TTF)$_4$[Hg$_3$I$_8$]$_{(1-x)}$ has a high transition temperature of $T_c = 8.1$ K. Therefore in a $V-T$ phase diagram it is proposed to be located very close to the charge order transition. Already slightest changes in the anion chain structure can push the system over the charge order transition. Then it becomes insulating at low temperatures instead of superconducting. The obtained spectra showed this effect. The Drude response disappeared and a charge order gap appears. The spectral weight was transferred to the charge fluctuation and MIR band. Known from pressure dependent DC measurements [134] a slight pressure of 300 bar can push the system back to the metallic side of the charge order transition and it becomes superconducting, again.

Comparing the spectra from the β''-(BEDT-TTF)$_2$SF$_5$CH$_2$CF$_2$SO$_3$ with other organic conductors and superconductors gives insight into the influence of the spectral weight evolution of the low frequency features. The isostructural metal was already compared in Fig. 10.27 in Sec. 10.2.3. The comparison with other organic conductors is given in Fig. 11.5 for the low temperature phase at 10 K. As a common feature in all system a transfer of spectral weight to the low frequency features takes place. That is due to the effective correlations. Important is how the spectral weight is distributed between the Drude response and the charge fluctuation band. The comparison of β''-(BEDT-TTF)$_2$-SF$_5$CH$_2$CF$_2$SO$_3$ with the isostructural β''-(BEDT-TTF)$_2$SF$_5$CHFSO$_3$ showed that the metallic properties, instead of superconducting, are due to the lack of interaction between the charge carriers. The correlations are too low to turn the effective interaction attractive. As a result the Drude spectral weight is strong, while the charge fluctuation feature is nearly absent.
Another way to reach a dominating metallic behavior is to change the filling to an incommensurate band filling like fifth-filling. This acts like carrier doping on the quarter-filled lattice. No complete charge order pattern can be formed anymore. There will be always a free lattice site with less occupied neighbors available to hop to. This results in a nearly pure Drude in the low frequency response as observed in the fifth-filled compound β''-(BEDO-TTF)$_5$(CsHg(SCN)$_4$)$_2$ in Fig. 11.5.

Within quarter-filled systems going from the metallic response like in the low correlated β''-(BEDT-TTF)$_2$SF$_5$CHFSO$_3$ to the superconducting state the correlations have to be

11. The Interplay of Charge Order and Superconductivity

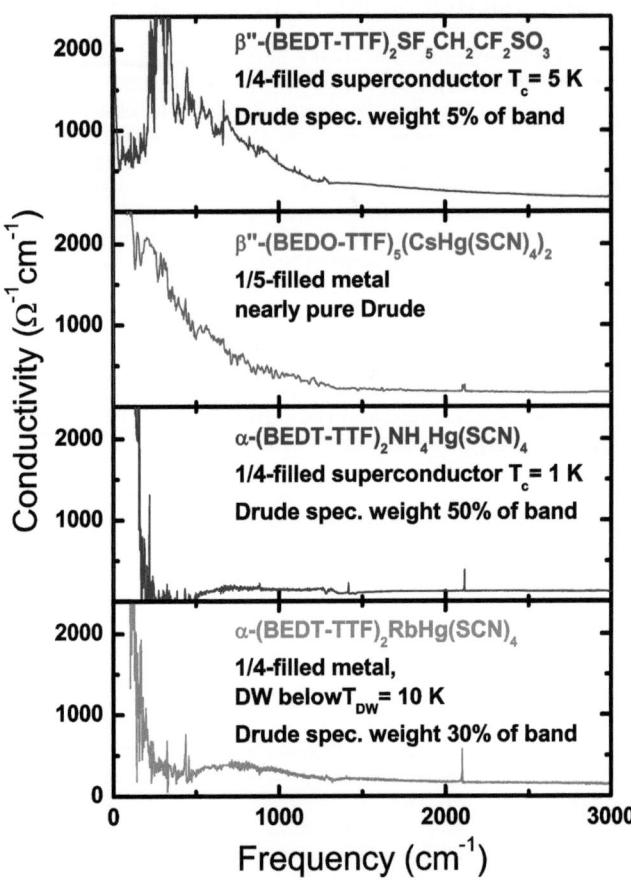

Figure 11.5.: Comparison of the optical conductivity at 10 K along the interstack direction (b-axis) in β''-(BEDT-TTF)$_2$SF$_5$CH$_2$CF$_2$SO$_3$ with several other organic conductors and superconductors. Additional data provided by N. Drichko.

increased to form an attractive interaction between the charge carriers. The stronger the attractive interaction, the higher should be the superconducting transition temperature. In α-(BEDT-TTF)$_2$NH$_4$Hg(SCN)$_4$ the presence of charge fluctuations is evident directly from the splitting of the charge sensitive modes in vibrational spectroscopy. The in-plane charge fluctuation band has been identified before [67, 78] as summarized in Sec. 2.2.4. Looking at the distribution of low frequency spectral weight one finds the Drude response to be fairly high. Its spectral weight is about 50% of the band spectral weight (Fig. 11.5). This explains the rather low $T_c \approx 1$ K. The system is located in the proximity to the charge order transition. There charge fluctuations turn the effective interaction between the charge carriers attractive. But compared to the β''-(BEDT-TTF)$_2$SF$_5$CH$_2$CF$_2$SO$_3$ which has a $T_c \approx 5$ K the α-(BEDT-TTF)$_2$NH$_4$Hg(SCN)$_4$ is located to the less correlated side. Also within the quarter-filled α-(BEDT-TTF)$_2$MHg(SCN)$_4$ family the α-(BEDT-TTF)$_2$NH$_4$Hg(SCN)$_4$ is estimated to be at low correlations. In comparison to that α-(BEDT-TTF)$_2$RbHg(SCN)$_4$ is found to be stronger correlated [78]. Indeed in the spectral weight transfer it is found that only 30% of the band spectral weight is present in the Drude response. But instead of becoming superconducting with a higher T_c, α-(BEDT-TTF)$_2$RbHg(SCN)$_4$ becomes insulating below $T = 10$ K due to structural influences. This is found also for other α-(BEDT-TTF)$_2$MHg(SCN)$_4$ family members. That structural changes can induce metal-insulator transitions is also reported for some of the systems that have been under investigation in this thesis (c.f. Sec. 4). For instance β''-(BEDT-TTF)$_2$SF$_5$CHFSO$_3$ can show a resistivity upturn at lowest temperature. As possible explanation a disorder of the anion is discussed similar to the process reported in β''-(BEDT-TTF)$_2$SF$_5$CHFCF$_2$SO$_3$ and described in the materials section. Similar structural changes are discussed as explanation for the superconductor-insulator transition in the pressure dependence of β''-(BEDT-TTF)$_2$SF$_5$CH$_2$CF$_2$SO$_3$. Even stronger structural changes lead to an ordered phase, as in β'-(BEDT-TTF)$_2$SF$_5$RSO$_3$ (R=CH$_2$, CF$_2$), e.g. Further, the influence of structural changes just by interacting with the anion layer is seen in β-(EDT-TTF)$_4$[Hg$_3$I$_8$]$_{(1-x)}$. It turns insulating at low temperatures because of a different anion configuration. Small deviations change the effective correlation. The superconductivity can be recovered under pressure. But for stronger deviation also there transitions to insulating states are found which can not be pushed back into a superconducting state [134].

11.3. Lattice phonon and collective charge order excitation

The presented work shows the close connection between the charge order and superconductivity in quarter-filled two-dimensional organic conductors and superconductors. The experimental findings can be nicely described with a pure electronic picture of nearest neighbor interaction driven charge fluctuation. Nevertheless, already the experiments show the strong influence of lattice dynamics. In particular these couplings are observed: A collective excitation of the charge order via a lattice phonon, strong e-mv coupled vibrations due to a modulation of the HOMO energies of the molecules, and a direct coupling to the charge on the molecule as seen in the charge sensitive molecular vibra-

11. The Interplay of Charge Order and Superconductivity

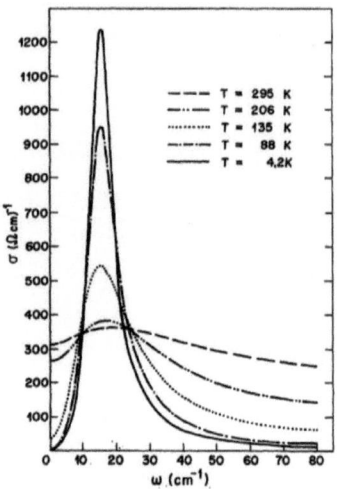

Figure 11.6.: Low frequency response along the chains of $K_2Pt(CN)_4Br_{0.3}$ 3.2 H_2O in the charge density wave state. The mode at 20 cm^{-1} is interpreted as collective response of the charge density. The pinning to the lattice and impurities shifts the collective mode to finite frqeuncies. From [192].

tions.

In the picture of collective excitations the feature observed in β''-(BEDT-TTF)$_2$-SF$_5$CH$_2$CF$_2$SO$_3$ is very strong due to a coupling of the lattice phonon to the strong electronic background. A similar phonon is observed in α-(BEDT-TTF)$_2$I$_3$ (cf. [61] and Fig. 2.15). There the mode is not strong. Also a temperature dependent study does not show an enhancement of the mode below the phase transition at 135 K [71]. The reason is that no electronic background is present in the charge ordered phase, which the phonon could couple to. In that sense the lattice phonon itself is not a unique feature. But the strong coupling to the electronic background in the β''-(BEDT-TTF)$_2$SF$_5$CH$_2$CF$_2$SO$_3$ makes it a collective excitation of the charge fluctuation. Its symmetry and detailed coupling mechanism is not known so far. Also the influence of this mode to the superconducting state is still unresolved. Its properties are comparable to a collective pinned mode resonance.

A prominent example for such a mode exists for $K_2Pt(CN)_4Br_{0.3}$ 3.2 H_2O (KCP) shown in Fig. 11.6. Its collective pinned mode resonance at 20 cm^{-1} is identified as the IR active phason excitation of the collective condensate response of the CDW state [192]. There the density wave state is incommensurate. For a commensurate density wave the

11.3. Lattice phonon and collective charge order excitation

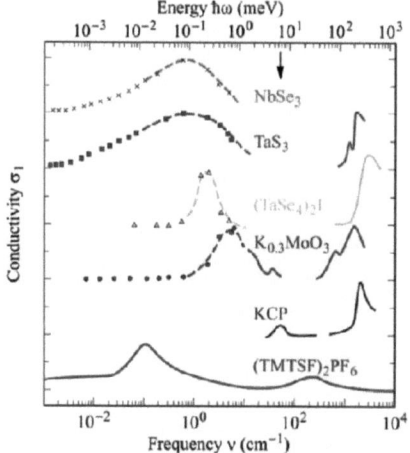

Figure 11.7.: Incommensurate pinned mode resonances within the single particle gap for a variety of materials. From [22].

pinning would be stronger and the collective mode should appear at higher frequencies. These collective excitations are a general feature. Several incommensurate pinned mode resonances, also including KCP are shown in Fig. 11.7 for different types of materials.

As collective excitation, the 40 cm^{-1} mode in the β''-(BEDT-TTF)$_2$SF$_5$CH$_2$CF$_2$SO$_3$ superconductor is similar to excitations interpreted as charge density wave excitations in manganites [193, 194] or nickelates [195], e.g. For Pr$_{0.7}$Ca$_{0.3}$MnO$_3$ the low frequency mode is shown in Fig. 11.8 which is assigned to a collective CDW excitation of the charge order pattern with orbital strips along the c-axis. However, that mode also persists above the charge order temperature [193]. In a series of manganites like in La$_{1/4}$Ca$_{3/4}$MnO$_4$, e.g., the collective mode excitation is nicely developed within the gap [194]. Fig. 11.9 shows the peaks assigned as the phason mode at 7.5 cm^{-1}. The two side bands are explained by the phason overtone and the also IR active phason-amplitudon combination mode. Also in manganites the stripe phase is phenomenologically similar to a sliding CDW compared by resistance hysteresis and broadband noise measurements by Cox et al. [196].

In the bad metal regime of La$_{2-x}$Sr$_x$CuO$_4$ (LSCO) low frequency modes are found [186, 198] which can be discussed as collective excitation of a stripe charge order and fluctuating stripes like in La$_{1.275}$Nd$_{0.6}$Sr$_{0.125}$CuO$_4$ (LNSCO) [197]. A comparision of LSCO and LNSCO is shown in Fig. 11.10. In LSCO the broad low frequency contributions are assigned to a collective mode describing the excitations of commensurate charge-stripes [186, 197, 198]. The strong enhancement in LNSCO is attributed to fluctu-

11. The Interplay of Charge Order and Superconductivity

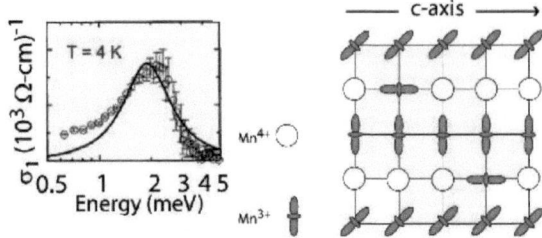

Figure 11.8.: Low frequency response of $Pr_{0.7}Ca_{0.3}MnO_3$. The 15 cm^{-1}/ mode is assigned to a collective CDW excitation. The charge and orbital ordering pattern is shown on the right. From [193].

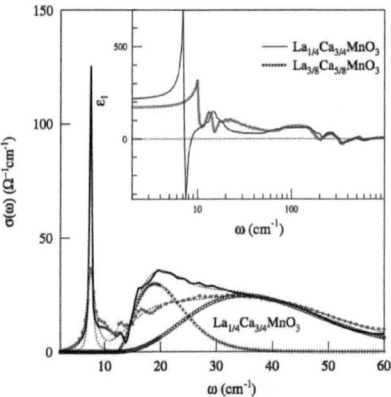

Figure 11.9.: Low frequency response of $La_{0.25}Ca_{0.75}MnO_3$ at 10 K (solid line) and 100 K (red dots). The modes are assigned to a collective CDW excitation: The phason at 7.5 cm^{-1}, its overtone at approximately 15 cm^{-1}, and the phason-amplitudon combination mode at 37.5 cm^{-1}. The inset shows the real part contributions to the dielectric function of the CDW at two diffrent doping levels. From [194].

11.3. Lattice phonon and collective charge order excitation

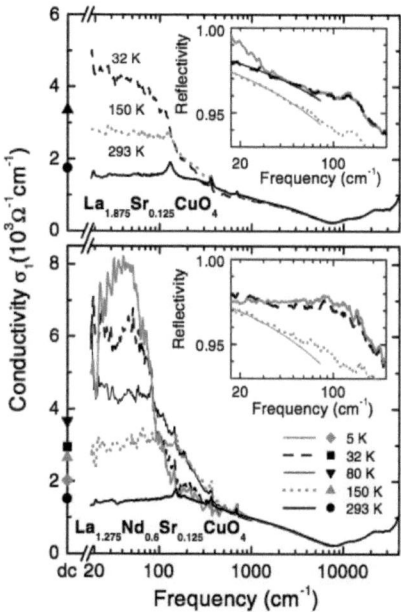

Figure 11.10.: Low frequency response of LSCO and the doped LNSCO. Besides the broad low frequency contribution in LSCO an additional peak below 100 cm^{-1} appears in LNSCO below 150 K and enhanced on decreasing temperature. From [197].

11. The Interplay of Charge Order and Superconductivity

ating stripes [197]. This is also observed for LSCO with incommensurate stripe patterns [198].

In θ-(BEDT-TTF)$_2$I$_3$, which was discussed before, the possibility of lattice or molecular vibrations as hidden excitations for the collective modes are considered instead of a pure electronic origin [191]. Further, low-frequency modes are present in the spin chain and spin ladder compound Sr$_{14}$Cu$_{24}$O$_{41}$: A mode at 1.8 cm^{-1} which so far is unexplained and at low temperatures a mode that shifts down from 14 to 11 cm^{-1} at the phase transition [199]. These might be connected to structural changes of the crystal lattice together with a CDW phase transition in the ladders, or with a charge order transition in the chains [200–202].

In the present case of β''-(BEDT-TTF)$_2$SF$_5$CH$_2$CF$_2$SO$_3$ we interpret the low frequency mode as result of the interplay of a lattice phonon with the charge order while the charge fluctuation band at higher frequencies shows a more electronic origin. But also for this band an alternative description exists in a full polaronic picture [178]. It considers a strong coupling of the electronic system to the lattice. But such a scenario would lead to stronger charge disproportionation in contrast to the measured values in the charge ordered metals. However the possibility of an enhancement of the electronic effect due to polaronic effects is likely. In summary one could argue that in the case of β''-(BEDT-TTF)$_2$SF$_5$CH$_2$CF$_2$SO$_3$ the charge fluctuation band is mainly due to electron-electron interactions and originates in the collective charge mode that drives the charge fluctuations. On the other hand the low frequency peak is a direct collective excitation of the charge order pattern in the system via a phonon mode. The fluctuating charge-order enhances the mode like in the case of incommensurate stripes in LSCO or the fluctuating stripes in LNSCO.

11.4. Vibrational influence to superconductivity

The possibility of polaronic enhancement effects, discussed for the charge fluctuation band, or coupling to charge-order, turn the interest to the phonons to enhance superconductivity. Besides that also the molecular vibration have a strong influence to the electronic system due to e-mv coupling. This coupling is significant as seen in the Fano shaped resonances of several modes of (B)EDT-TTF. The electron-phonon vortex in the the charge ordered metallic state is strongly renormalized due to charge fluctuations. Then, in vicinity to the charge order transition, also the molecular vibrations could contribute to superconductivity even with anomalous pairing (d_{xy}). This could solve the restriction of the pure electronic picture of charge fluctuations mediating the superconductivity which can explain low T_c's, only. As seen in Fig. 3.8 the effective coupling in the simple electronic case $\lambda_{d_{xy}} < -0.01$ is very small. Assuming this λ it is not possible to understand a T_c in the order of several Kelvin. Molecular vibrations are a good candidate to overcome these limitations due to their strong coupling to the electronic system. Intramolecular vibrations can be included as Holstein-like phonons into the extended Hubbard model. Coupling parameters for this model are calculated in Fig. 11.11. It shows the e-e, e-ph, and total coupling for s- and d_{xy}-symmetry for two different on-site

11.4. Vibrational influence to superconductivity

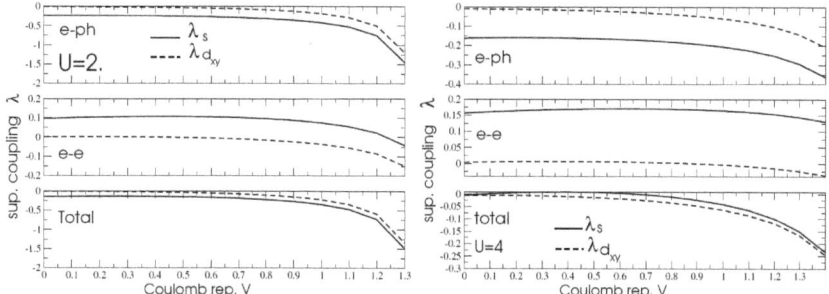

Figure 11.11.: Electron-phonon, electron-electron, and total coupling parameters for Holstein-like intramolecular vibrations implemented to the extended Hubbard model for different on-site repulsion: (left) $U = 2$ and (right) $U = 4$. As bare non-renormalized value $\lambda = 0.4$ is assumed [98, 111, 203]. Calculation from A. Greco.

correlations U. When V approaches the critical value V_c the el-ph coupling increases even in the d_{xy}-channel. Further the λ values are much larger for phonons than for the pure electronic case. It is also interesting to notice that for small V there is no possibility for superconductivity. There the system starts with an isotropic and attractive e-ph $\lambda \sim -0.4$ (top) but the correlations (middle) kill the effect. Close to charge order, the vibrations develop an attractive interaction in both the s and the d_{xy} channel which is not compensated by correlations. In the d_{xy} channel the correlations give even a positive contribution to superconductivity. For higher U the e-e correlations are more dominating since e-ph coupling is reduced. In total the d_{xy} channel is favored. At small U the s-wave channel should be more favored and lead to higher attractive interaction λ: The on-site screening is reduced. That is in agreement with the experiments which show a rather high T_c of several Kelvin and the opening of a BCS-like gap without fingerprints for nodes in agreement with s-wave symmetry.

Part V.

Summary and Outlook

12. Summary

This thesis is a comprehensive optical study of the interplay of charge order and superconductivity in two-dimensional organic conductors. Reflectivity measurements have been performed in a broad frequency (8-20000 cm^{-1}) and temperature range (1.8-300 K). This allows us to obtain the optical response of organic molecular metals and superconductors. A peculiar interplay of charge-order and superconductivity is observed in the optical properties and discussed in a picture of correlation driven charge fluctuations. In addition mw- and DC-transport properties characterize the ground states of the correlated metals, charge-ordered insulators, and superconductors. IR vibrational and Raman spectroscopy allow to directly measure the charge distribution on the molecular sites. These help to characterize the electronic state and to compare the onset of charge-order or charge fluctuations with the appearance of optical features.

As prime examples to explore the interaction between the charge-ordered and superconducting state, the quasi two-dimensional quarter-filled organic superconductor β''-(BEDT-TTF)$_2$SF$_5$CH$_2$CF$_2$SO$_3$ and its isostructural sister compound β''-(BEDT-TTF)$_2$SF$_5$CHFSO$_3$ are in the focus of the experimental investigations. These two systems differ from each other in the degree of effective electronic correlations, only. Therefore they are ideal candidates to prove the correlation influence on the ground states. The optical response reveals four main features (Fig. 11.2). (i) A broad charge transfer band in the MIR that characterizes the localization of charge carriers due to the electronic onsite (U) and intersite (V) correlations. In the pseudogap at frequencies below that band (ii) a small Drude peak represents the coherent charge carriers response. It is present at all temperatures. In a simple competition between the charge-ordered and the metallic state these two features would characterize the transition. However, here an interaction between the localized and coherent charge carriers is observed. An additional low frequency feature appears: (iii) the charge fluctuation band. It indicates the close relationship between charge order and superconductivity. A simple coexistence of these states and phase separation could not explain the additional band. Its basic behavior is similar to the low-frequency response known from bad metallic states of high-T_c's or other organic conductors. In addition, in the charge fluctuation regime (iv) a direct collective excitation of the charge-order in the system is observed via a low frequency phonon mode. This is very similar to collective excitations in other correlated materials like cuprates, manganites, etc., but also to charge order resonances, in general.

Here the microscopic origin of the systems electronic behavior including its collective excitations can be compared to a charge fluctuation picture drawn in a theoretical approach. Within this model it is possible to describe the nature of these excitations and

12. Summary

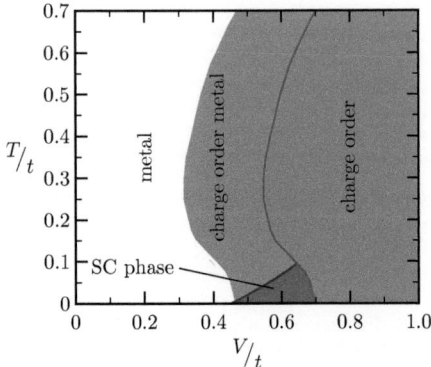

Figure 12.1.: Proposed phase diagram for quasi two-dimensional quarter-filled systems of an extended Hubbard model on a square lattice. Calculated for $U = 20t$. Close to the charge order transition traces of charge fluctuation are present. The behavior can be explained by a charge ordered metallic state. From [17].

the intriguing physics leading to superconductivity in the two-dimensional organic conductors at quarter-filling. The interplay taking place can be summed up in the phase diagram shown in Fig. 12.1. For $V = 0$, the uncorrelated case, the system behaves purely metallic. At finite but low V the correlations lead to the onset of charge fluctuations. They renormalize the metallic state and destabilize it towards a quantum phase transition. At a critical correlation value (red line) the system undergoes a charge-order transition into an insulating phase at high correlations. Of special interest is the area in the proximity of the phase transition but still on the metallic side (shaded green). In that region a charge-ordered metallic state is proposed which shows a strong interaction between the metallic phase and the charge fluctuations. The fluctuations can act as attractive effective potential for the charge carriers and allow the formation of Cooper pairs in a superconducting state (blue). Therefore superconductors have to be highly correlated systems close to the charge order transition. For high temperatures they are expected to be in a charge-ordered metallic regime. For the superconductors investigated here the presence of this state is proven. Metallic transport properties and at the same time charge fluctuations are found. The dynamical properties are characterized by optical spectroscopy and are discussed within the interaction picture of quasi-free charge carriers and localized carriers via charge fluctuations.

The transport properties are measured using standard DC- but also mw-techniques. The superconductors show a perfect metallic behavior down to the superconducting transition. In the mw-properties right above the superconducting transition a resistivity upturn in β''-(BEDT-TTF)$_2$SF$_5$CH$_2$CF$_2$SO$_3$ is observed as trace of charge fluctuations. The same samples are analyzed in their vibrational properties using IR vibrational spectroscopy and Raman. Therefore a low temperature stage for a micro cryostat is developed to enable vibrational spectroscopy at low temperatures under a microscope. For the organic superconductors a small but distinct splitting in the charge sensitive modes of the BEDT-TTF molecules directly proves a charge disproportionation between the molecular sites. On cooling the amount of charge disproportionation does not change. For the less correlated metals no significant splitting is found; in average the charges are equally distributed among the molecular sites. For charge-ordered insulating systems a large splitting of the charge sensitive modes is observed indicating a large charge disproportionation between the sites which is found to set in directly at the phase transition without a gradual increase of disproportionation.

The dynamical properties of the electronic states are investigated via in-plane IR spectroscopy. The optical response shows the properties of the localized carriers in a MIR charge transfer band while the quasi-free charge carriers give rise to a Drude response. For the β''-(BEDT-TTF)$_2$SF$_5$CH$_2$CF$_2$SO$_3$ superconductor most of the spectral weight is in the localized part of the spectrum. The Drude response is very small and contains below 5% of the total band spectral weight. An extended Drude analysis of the coherent carrier response shows a linear frequency dependence in the scattering rate. That reveals an interaction with a bosonic excitation. In the present case these are the correlation driven charge fluctuations. Due to that the charge fluctuation band arises at low but finite frequencies. That is similar to the low frequency band in bad metals. It is found to be of mainly electronic origin due to the charge carriers interacting with the charge fluctuations. The band grows on cooling in parallel to the electronic background of the Drude peak. That also shows that the interaction is based on an interplay between the metallic and localized carriers and not on a competition.

Besides this electronic feature also an electron-phonon coupled collective response is found with the onset of charge order fluctuations. It shows up as peak at a frequency even below the charge fluctuation band. Its origin is explained as lattice phonon that directly excites the differently charged sites against each other: a collective charge-order excitation.

Comparing these features to the metallic β''-(BEDT-TTF)$_2$SF$_5$CHFSO$_3$, only weak traces of a small charge fluctuations are found. Most spectral weight is located in the Drude response. That shows the less interaction of the ordered and metallic states in this compound. The metal character is dominating and due to the absence of charge fluctuations the system does not become superconducting.

In the superconducting state below 5.4 K an optical gap opens in the β''-(BEDT-TTF)$_2$-SF$_5$CH$_2$CF$_2$SO$_3$. The temperature dependent opening of the gap can be described by a BCS fit for arbitrary purities. The gap frequency can be estimated to about $2\Delta = 12$ cm^{-1}. The $2\Delta/(k_b T_c) = 3.3$ ratio is close to BCS expectations for a weak coupling and $2\Delta/\tau^{-1} \sim 1$ puts it to the intermediate purity regime. Basically the narrow Drude response is gapped. Applying a magnetic field destroys superconductivity and closes the gap. That proves, that indeed the observation is due to superconductivity.

12. Summary

Comparing the results from the β''-(BEDT-TTF)$_2$SF$_5$CH$_2$CF$_2$SO$_3$ superconductor and the isostructural β''-(BEDT-TTF)$_2$SF$_5$CHFSO$_3$ with other quarter-filled systems qualitatively allows us to connect the correlation strength to the strength of the interaction features and to the superconducting temperature (Fig. 11.5). If the fluctuations are absent or weak, the system stays metallic. If a critical correlation value is reached the charge fluctuations can mediate an attractive interaction and the system turns superconducting. Comparing the strength of the fluctuation band to the metallic Drude response shows that the closer the system is to the charge order transition, the stronger are the interactions. In that case the critical temperature increases. The less correlated system β''-(BEDT-TTF)$_2$SF$_5$CHFSO$_3$ stays metallic, the α-(BEDT-TTF)$_2$NH$_4$Hg(SCN)$_4$ with still a strong Drude response turns superconducting around $T_c = 1$ K. The stronger correlated β''-(BEDT-TTF)$_2$SF$_5$CH$_2$CF$_2$SO$_3$ has only a very small Drude but a strong charge fluctuation band and becomes superconducting at around $T_c = 5.4$ K. The β-(EDT-TTF)$_4$[Hg$_3$I$_8$]$_{(1-x)}$ is so close to the charge order transition that even smallest deviations in the anion chain stoichiometry can turn the system insulating. The superconducting transition there is at $T_c = 8.1$ K.

Tiny changes in the anion layer or other small structural changes can alter the physical properties significantly. That is seen and well understood in the (BEDT-TTF)$_2$SF$_5$$RSO_3$ family. Also in the α-(BEDT-TTF)$_2$$M$Hg(SCN)$_4$ family these influences are present. Located in the proximity of the charge order transition the NH$_4$-salt is the only one that becomes superconducting. The others like the Tl-salt also show distinct charge disproportionation in their vibrational properties and a charge fluctuation peak present in the in-plane optics. Some like the Rb-salt are even stronger correlated. But they becoming insulating at low temperatures due to structural effects. A strong coupling of the electronic system to the lattice is seen e.g. in θ-(BEDT-TTF)$_2$RbZn(SCN)$_4$. There the charge-order transition is accompanied by a structural transition. From the electronic point of view the θ-phase is similar to the β-phase. However, at the charge-order transition the molecules in the conducting plane rotate against each other. That leads to a doubling of the unit cell which makes the structure similar to the one of α-(BEDT-TTF)$_2$I$_3$. The phase transition into the charge-ordered insulating phase in both systems looks similar as traced by vibrational spectroscopy. So the driving parameter for the structural changes remains a puzzling question. No direct connections to the electronic phase transition could be found. Compared to the charge-ordered metallic state of the superconductors the insulating charge-order shows a large charge disproportionation between the molecular sites. Entering the phase transition this charge disproportionation sets in in one step. No gradual increase of the disproportionation is found.

Besides the structural arrangement, the structural dynamics directly influences the electronic state. The collective excitation of the charge order shows its strong enhancement due to coupling of the lattice phonon to the electronic background. Another direct influence to the electronic system is seen in the e-mv coupled modes of the molecular vibrations. Also the charge fluctuation band is assumed to be enhanced by plasmonic effects, which could help to explain the rather high transition temperatures realized in the organic superconductors. Close to charge-order the electron-phonon coupling is strongly renormalized due to the fluctuations and the vibrational modes can contribute to superconductivity. First calculations taking the coupled vibrational modes into account show that the superconducting coupling constant increases significantly in amplitude and stronger attractive interactions are possible.

Summarizing, this thesis shows the close connection between the charge-ordered state and superconductivity in the quarter filled two-dimensional organic conductors. Superconductivity can be mediated by charge fluctuations. The basic features can be explained by an interplay via charge fluctuations in the metallic state. In their general behavior they can be compared to collective excitations of electronic or phononic origin, which are also present in other correlated systems. Deviations can be explained by structural changes or the influence of the lattice dynamics, which turns out to be of importance for the electronic properties. First estimations point out that intramolecular vibrations can have a significant contribution to superconductivity.

13. Outlook

First attempts to extend the picture based on the results here are already motivated in the discussion. The inclusion of vibrational features and phonons is a promising way to investigate new interactions and contributions to superconductivity. Therefore a more detailed understanding of the lattice dynamics in general is important. The different coupling of lattice phonons and vibrational features has to be taken into account and their interplay with the correlations has to be investigated. Besides detailed investigations of the lattice system in thermal equilibrium also the time scales on which the electronic or vibrational system react become important. First experiments on photoinduced melting of charge order report of a instant electronic response and a slow response of lattice system which follows [184, 204]. The importance of these slow driven oscillations is seen in their report on charge order stabilizing and destabilizing properties of different coherent phonon modes [184, 204, 205]. To explore the interplay of the electronic and lattice system in that detail, time resolved methods are needed to decouple the electronic and phononic time scales.In addition the lattice dynamics might directly induce electronic changes by direct coupling to the charge fluctuations e.g via the collective modes found in β''-$(BEDT-TTF)_2SF_5CH_2CF_2SO_3$. Also the electronic correlations are possibly controllable via exited vibrational states of the molecules.

Applying pressure is also an excellent tool to change the effective correlations [121, 130–132, 147]. Following the changes in optics should reveal more metallic properties with increased pressure. High correlated, charge-ordered systems could be pushed back into the metallic phase and at low temperatures even become superconducting. The correlation tuning can be done more gradual than by chemical pressure. Then the phase transition can be investigated as function of effective correlation and compared to the transition as function of temperature.

Another tuning possibility of the systems is to change the potential by applying an external field and therefore current. That changes the internal correlations or break the symmetry in charge-ordered state [87]. Since internal fields will build up against the external ones it is difficult to explore the local effect of the fields. However, symmetry breaking effects should not be affected by this problem. A more promising way to apply external fields is to change the effective filling via field effects [85]. Interesting perspectives to investigate are different responses of systems with commensurate or incommensurate filling and how the pictures of the electron dynamics in e.g. half-filled and quarter-filled systems gradually shift into each other. Another question is the influence of non-stoichiometric filling which acts as slight doping with charge carriers.

Exploring the dependency of correlations and filling of the system are important steps

13. Outlook

to a unified view and towards a general phase diagram for correlated two-dimensional organic systems. Based on the structure of the systems, the electronic as well as the phononic system shows significant contributions. They are strongly connected in an intriguing interplay. Both have to be included into the generalized picture.

Further the comparison to other correlated systems is a topic of major interest. A lot of characteristic features show up in a similar way for different classes of materials. To some extend they can be described by the same phenomenological description. However, the microscopic picture and the detailed physics of the systems are completely different. The point of interest in that case is what concepts are of general importance to describe the properties of correlated electron systems and how are they translated into the detailed model describing the actual physics of the investigated system. Examples are the collective modes, which phenomenological describe the bad metal character of correlated systems. An example investigated here is the charge fluctuation band. It is identified to depend on electron-electron interactions, which drive the phase transition in the quasi two-dimensional quarter-filled organic superconductors. An additional collective mode is the direct excitation of charge-order via a lattice phonon which shows the importance of electron-phonon coupling. Even being of different microscopic origin both modes can be described in general as collective excitation or depinning of a charge order resonances.

Part VI.

Appendix and Bibliography

14. Appendix

14.1. Slave boson treatment on the extended Hubbard model

The nearest neighbor interaction driven charge fluctuation model described in Sec. is based on an extended Hubbard model $H = T + U + V$. For very high on-site Coulomb repulsions U double occupancy of the molecular sites are very unlikely, so it is possible to treat the system in the large U limit. Tuning $U \to \infty$ only single occupancy is allowed. In this case the Hamiltonian (3.1) reduces to the Hamiltonian of the t-V model

$$H = t \sum_{\langle ij\rangle,\sigma} P\left(c_{i\sigma}^{\dagger} c_{j\sigma} + c_{j\sigma}^{\dagger} c_{i\sigma}\right) P + V \sum_{\langle ij\rangle} n_i n_j - \mu \sum_{i\sigma} n_{i\sigma} \qquad (14.1)$$

where the P operator projects out the double occupied states. This $U \to \infty$ model can then be analyzed within the slave-boson theory [206–208] where the electron operator is replaced by the product of a spinless boson operator b_i representing the charge, in this case an empty site, and a neutral fermion operator $f_{i\sigma}^{\dagger}$ carrying the spin ($c_{i\sigma}^{\dagger} = f_{i\sigma}^{\dagger} b_{i\sigma}$). Using the SU($N$) generalization where the spin index σ is extended to rum from 1 to N, McKenzie et al. [44] worked on the extended Hubbard model on a square lattice at quarter filling to describe the rich physics of the θ–type organic conductors. They also show the relevance of this theory for the β''–systems on a triangular lattice.

In large-N the local constraint can be written as $f_{i\sigma}^{\dagger} f_{i\sigma} + b_i^{\dagger} b_i = N/2$ where in the relevant case of N=2 (spin $\pm\frac{1}{2}$) this means that the site is not allowed to be double occupied. This constrain is later taken into account by a Lagrange multiplier λ controlling the non-double occupancies criteria. This treatment shows that the whole physics can be described by the charges b and their distribution λ. The relevance of the large-N approach relays on the fact that it provides a controllable approximation in power of the small parameter $1/N$.

14.1.1. Nearest neighbor interaction driven charge order

Within the mean field solution the boson fields are taken by their mean field values $b = \langle b_i\rangle$ and $\lambda = \langle\lambda_i\rangle$. The solution describes a renormalized Fermi liquid with the eigenenergies given by

$$\epsilon_{\mathbf{k}} = \frac{-tb^2}{N} T_{\mathbf{k}} + \lambda - \mu + 4V\frac{n}{N} \qquad (14.2)$$

where $T_{\mathbf{k}} = 2(\cos k_x + \cos K_y)$ is the Fourier transform of the kinetic energy T, describing the band, in the units of nearest neighbor hopping t. So the spatially homogeneous and

14. Appendix

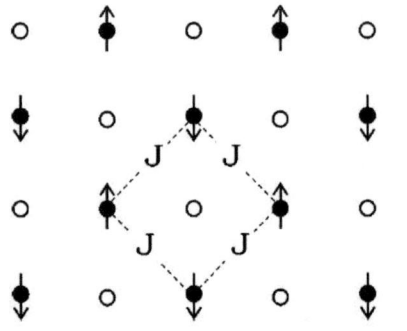

Figure 14.1.: Charge fluctuation induced checker board or diagonal stripe like charge order pattern on the 2D square lattice due to divergence of the charge susceptibility at the (π, π)-wavevector. From [44].

time independent boson fields b^2 and λ describe the band renormalization and a band shift while the nearest neighbor interaction V leads to a shift in the chemical potential. The minimization conditions of the free energy for b reads as $b^2 = N/2 - n$. For $N = 2$ in the case of quarter filling, $n = \frac{1}{2}$, the bandwidth is reduced to a half and so the effective mass increases by $m^*/m = 1/b^2 = 2$.

The fluctuations about this mean field solution were introduced by dynamic fluctuations around the mean field boson fields $r = b + b\delta r(\tau)$ and $-i\lambda(\tau) = \lambda + i\delta\lambda(\tau)$. Here δr describes the fluctuations in the local charge density while $\delta\lambda$ represents the non-double occupancy constraint fluctuations, meaning really an fluctuation of the local charge arrangement. Effectively this results in a fermion boson Hamiltonian

$$H = H^f + H^b + H^{f-b} \tag{14.3}$$

where the fermionic part H^f describes the mean field solution, the bosonic part H^b the dynamical and spacial fluctuations about the mean field solution, and H^{f-b} the coupling between them. The fluctuations are shown to be driven by the nearest neighbor interaction V and destabilize the Fermi liquid for correlations above a critical value $(V/t)_c$ for a quasiparticle wavevector (π, π). There the charge susceptibility diverges and a quantum phase transition into a checker-board charge order occurs (Fig. 14.1). The critical values reads $(V/t)_c = 0.78$ for quarter filled systems with a constant density of states and $(V/t)_c = 0.69$ for the density of states on a square lattice. The influence of next nearest neighbor hopping along one diagonal t' increases the critical value with increasing diagonal hopping up to $(V/t)_c = 0.95$ for the triangular lattice $(t'/t = 1)$ as given in Fig 14.2. That means for the same value of correlations the system gets more metallic the more anisotropic the lattice. The charge order taking place is not due to a

14.1. Slave boson treatment on the extended Hubbard model

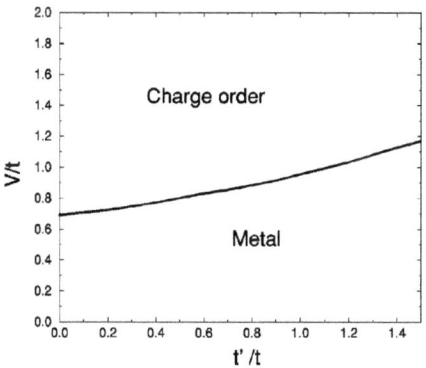

Figure 14.2.: $t - t'$-dependence of the critical value of the charge order transition for the 2D square lattice with next nearest neighbor hopping along one diagonal. An antiferromagnetic exchange interaction J, turned 90 degrees to the original lattice, appears between the sites in the charge ordered phase. From [44].

nesting of the Fermi surface, which is metallic as for a quarter filled square lattice as seen in Fig. 14.3(left), but purely due to the nearest neighbor electron-electron interactions. Since the dependence of the critical value $(V/t)_c$ is only weak it is justified to treat the pure square lattice as first approximation for a triangular lattice as well. That makes the model also applicable for β''-systems which are in the focus of our experimental studies. From the Fermi surface of the triangular lattice it is even clearer that the CO is due to correlation effects because nesting conditions for (π, π) are even more unlikely to fulfill (Fig. 14.3(right)). The checker board like charge order pattern (Fig. 14.1) results in an fourth order antiferromagnetic exchange interaction $J = \frac{4t^4}{9V^3}$ along the diagonals of the original square lattice for $V \gg t$. To exclude artifacts due to the approximations (mainly N=2 and finite U) Calandra et al. [69] performed exact diagonalization at zero temperature to show that the intersite Coulomb repulsion drives the metal insulator transition. Therefore they did Lancos calculation on $L = 8$, 16, and 20 sites clusters. Evaluating the Drude weight as function of intersite interaction (V/t) for different onsite interactions (U/t). Finite size effects were checked by comparing different cluster sizes L. With increasing nearest neighbor interaction the Drude weight decreases. At a critical value $V_c^{MI} \approx 2.2$ (for L=16) a metal to insulator transition occurs. Insulating behavior is checked via exponential drop of spectral weight with linear system size $\sqrt{L}$. The onset of charge order is seen in the charge correlation function. This function indicates checker board like CO. The critical values are around $V_c^{CO} \approx (1.4 - 1.8)t$ (at $U = 10t$). Decreasing U leads to an increase of this critical value. The CO starts to set in at the CO transition and can be described by a charge ordering parameter η. For $\eta = 0$ no

14. Appendix

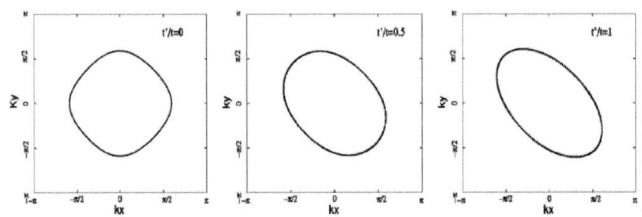

Figure 14.3.: Evolution of the Fermi surface from a square lattice (left) to a fully triangular lattice (right). In all cases no charge order due to nesting with (π, π) wavevector possible. From [44].

CO is established and charges are homogeneously distributed. $\eta = 1$ means fully formed CO. At the metal insulator transition the CO pattern is close to be complete ($\eta \approx 0.75$). Starting from there hopping of a charge within the CO would cost more than the kinetic hopping energy so localization sets in. Thats why a metallic CO state can be realized for $V_c^{CO} < V < V_c^{MI}$. In terms of spectral weight a mid-infrared band could be identified to be dominant in the metallic as in the CO regime. Close to the transition a transfer of spectral weight from the Drude and the Hubbard band at higher frequencies is transfered to it. The more detailed analysis of dynamical properties is given in Sec. 3.2.

At this point it is important to mention that the large-N approach (Ref. [18]) is free from salve particles and works on the basis of the original constrained fermions. At $N = \infty$ results of Ref. [18] agree with those obtained in the slave boson framework (Ref.[Merino and McKenzie]). Beyond mean field the method of Ref. [18] is useful for the calculation of self-energy effects.

14.2. Construction sketches of the sample holder

In the following the construction sketches of the sample holder for the micro cryostat are shown. First the mounted unit and then the single pieces including two alignment tools. The material of the stage including the positioning screws is brass. The ball fixing the position of the inner plate is stainless steel. Also the mounting screws for the outer ring and the inner plate are made of steel.

The functional principle is as follows: The outer ring is hold by the two arms at the side of the base plate. It allows tilting around one axis. It is pulled by a spring between the base plate and the outer ring at the one side while it is pressed by a nut from top at the other side. The position of the nut is adjustable in hight. In that way the degree of tilting of the outer ring is fixed.

The inner plate is fixed to the outer ring, therefore it has the same tilt as the outer ring around this axis. But it can be tilted around an axis perpendicular it. To allow

14.2. Construction sketches of the sample holder

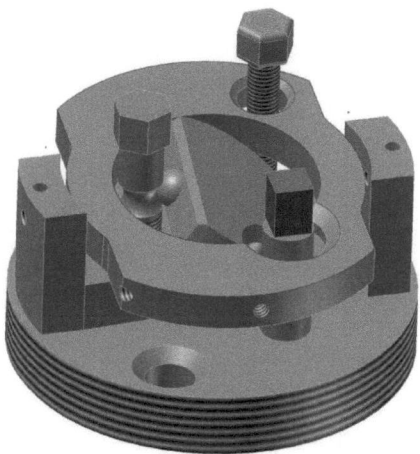

Figure 14.4.: The mounted group of the sample holder. The holder contains of a base plate, an outer ring, a inner plate and a positioning screw for the inner plate. The other ring positioning is done via a spring pulling from below and a nut pushing from top.

an independent alignment of the two axes, the tilt of the inner plate has to be fixed decoupled from the position of the outer ring. That is realized by the inner rod right on the outer tilting axes, next to the inner plate. It has a grove to hold to which a steel ball is pressed from the inner plate. The ball allows a smooth turning around the axis when tilting the other ring. It decouples the position of the grove independent of the outer tilt. The grove hight can be adjusted by screwing the rod in or out. Its position fixes tilt of the inner plate.
In that way an uneven shaped sample surface can be aligned parallel to the surface of a reference mirror. The flat mirror is placed on a fixed rod on the base plate.

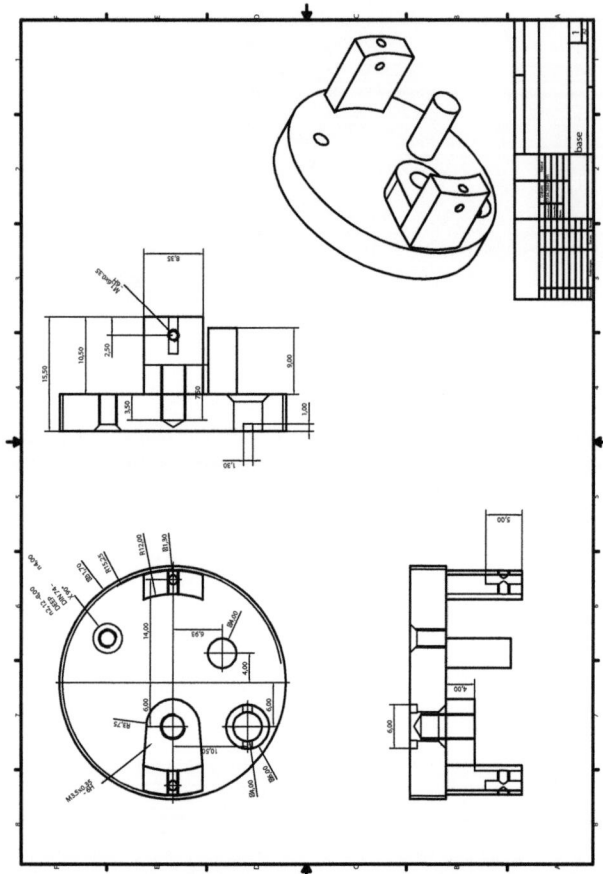

Figure 14.5.: The base plate with holders for the outer ring, and the mirror post. Mounting positions for spring and screw to align the ring tilting via pulling and pushing, respectively. The deeper thread is for the positioning screw of the inner plate.

14.2. Construction sketches of the sample holder

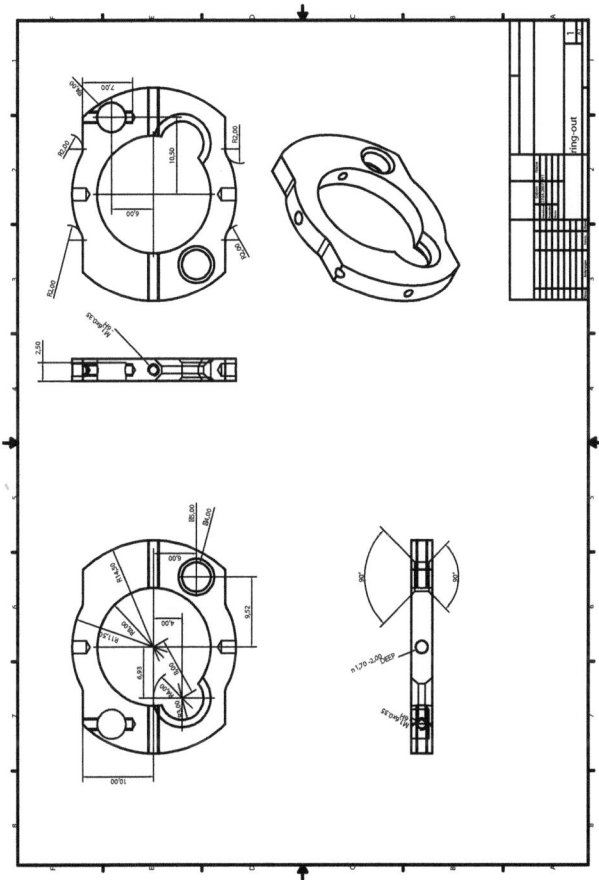

Figure 14.6.: The outer ring with mounts to the base plate and holders for the inner plate. Holes for spring and screw fixing the ring tilt.

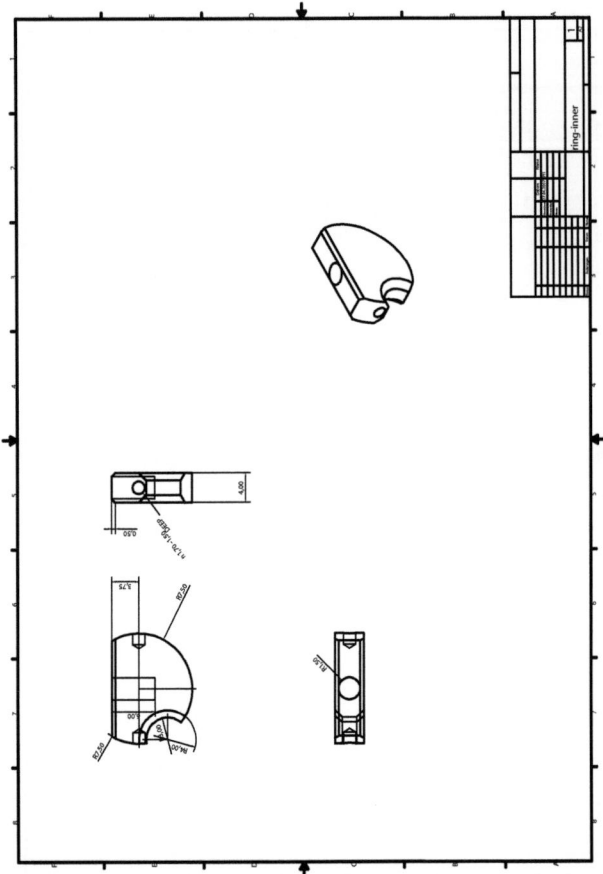

Figure 14.7.: The inner plate. Mounts to the outer ring. Circle segment spared for the positioning screw and the mirror post. On the side recess to hold a steel ball to be pressed against the positioning screw and fixing the tilt of the inner plate.

14.2. Construction sketches of the sample holder

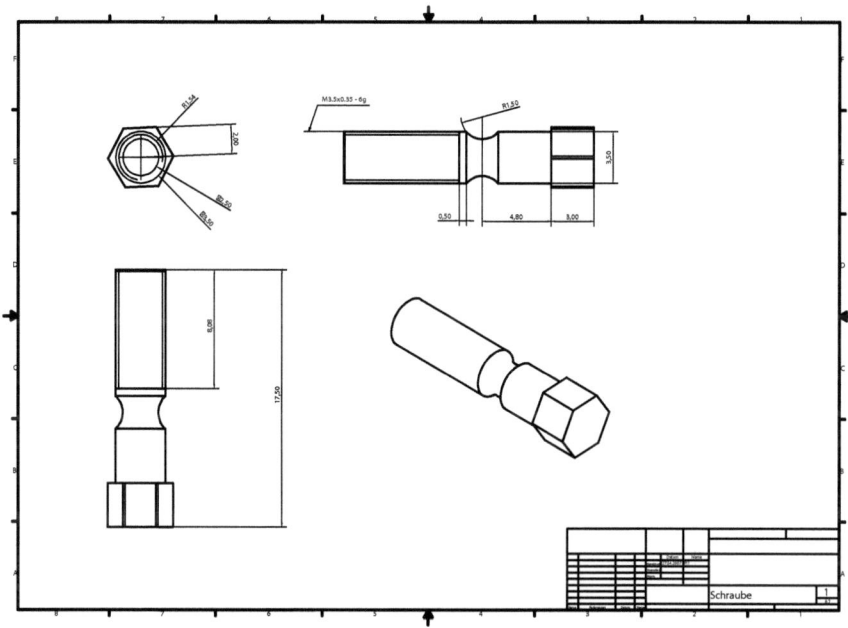

Figure 14.8.: Positioning screw. Carve to fix the position of the steel ball to adjust tilt of the inner plate.

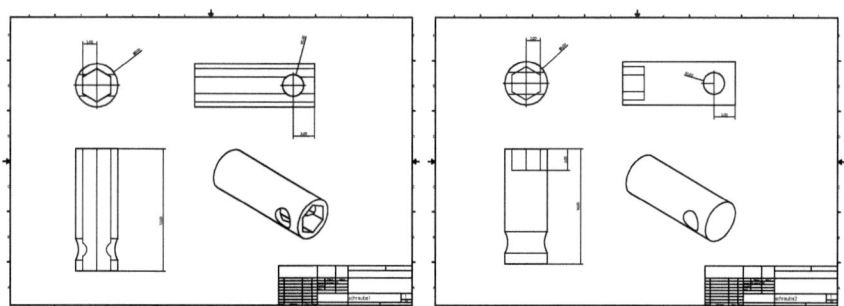

Figure 14.9.: Alignment tools to fix hight of the inner positioning screw and the nut pressing on the other ring.

Bibliography

[1] T. Ishiguro, K. Yamaji, and G. Saito, *Organic Superconductors*, 2nd ed., Springer series in solid-state sciences, Vol. **88** (Springer-Verlag Berlin Heidelberg New York, 1998).

[2] N. Toyota, M. Lang, and J. Müller, *Low-Dimensional Molecular Metals*, Springer series in solid-state sciences, Vol. **154** (Springer-Verlag Berlin Heidelberg, 2007).

[3] "Special issue: Molecular Conductors," Chemical Review **104** (2004).

[4] Z Fisk, DW Hess, CJ Pethick, D Pines, JL Smith, JD Thompson, and JO Willis, "Heavy-electron metals - new highly correlated states of matter," Science **239**, 33 (1988).

[5] J Orenstein and AJ Millis, "Advances in the physics of high-temperature superconductivity," Science **288**, 468 (2000).

[6] PW Anderson, "The Resonating Valence Bond State in La_2CuO_4 and Superconductivity," Science **235**, 1196 (1987).

[7] IG Austin and NF Mott, "Metallic and Nonmetallic Behavior in Transition Metal Oxides," Science **168**, 71 (1970).

[8] RH McKenzie, "Condensed matter physics - Similarities between organic and cuprate superconductors," Science **278**, 820 (1997).

[9] ND Mathur, FM Grosche, SR Julian, IR Walker, DM Freye, RKW Haselwimmer, and GG Lonzarich, "Magnetically mediated superconductivity in heavy fermion compounds," Nature **394**, 39 (1998).

[10] G Kotliar and D Vollhardt, "Strongly correlated materials: Insights from dynamical mean-field theory," Physics Today **57**, 53 (2004).

[11] K Bechgaard, CS Jacobsen, K Mortensen, HJ Pedersen, and N Thorup, "Properties of 5 Highly Conducting Salts - $(TMTSF)_2X, X=PF_6-, ASF_6-, BF_4-$ and NO_3-, Derived from Tetramethyltetraslenafulvalene (TMTSF)," Solid State Communications **33**, 1119 (1980).

[12] D Jerome, M Ribault, and K Bechgaard, "Organic superconductors," New Scientist **87**, 104 (1980).

[13] S Flandoris, C Coulon, P Delhaes, D Chasseau, C Hauw, J Gaultier, JM Fabre, and L Giral, "In the structure and properties of TMTTF and TMTSF salts - experimental-evidence for the importance of interchain couplings," Molecular Crystals and Liquid Crystals **79**, 307 (1982).

[14] D Jerome, "The development of organic conductors: Organic superconductors," Solid State Sciences **10**, 1692 (2008).

[15] H Seo, "Charge ordering in organic ET compounds," Journal of the Physical Society of Japan **69**, 805 (2000).

[16] H Seo, C Hotta, and H Fukuyama, "Toward systematic understanding of diversity of electronic properties in low-dimensional molecular solids," Chemical Reviews **104**, 5005 (2004).

[17] J Merino and RH McKenzie, "Superconductivity mediated by charge fluctuations in layered molecular crystals," Physical Review Letters **87**, 237002 (2001).

[18] J Merino, A Greco, RH McKenzie, and M Calandra, "Dynamical properties of a strongly correlated model for quarter-filled layered organic molecular crystals," Physical Review B **68**, 245121 (2003).

[19] A Greco, J Merino, A Foussats, and RH McKenzie, "Spin exchange and superconductivity in a t-J(')-V model for two-dimensional quarter-filled systems," Physical Review B **71**, 144502 (2005).

[20] P Drude, "On the electron theory of metals," Annalen Der Physik **1**, 566 (1900).

[21] L Degiorgi, "The electrodynamic response of heavy-electron compounds," Reviews of Modern Physics **71**, 687 (1999).

[22] Martin Dressel and George Grüner, *Electrodynamics of Solids* (Cambridge University Press, 2002).

[23] DN Basov and T Timusk, "Electrodynamics of high-T_c superconductors," Reviews of Modern Physics **77**, 721 (2005).

[24] S. V. Dordevic and D. N. Basov, "Electrodynamics of correlated electron matter," Annalen Der Physik **15**, 545 (2006).

[25] L. Degiorgi, "The drude model in correlated systems," Annalen Der Physik **15**, 571 (2006).

[26] LD Landau, "On the Theory of the Fermi Liquid," Soviet Physics Jetp-Ussr **8**, 70 (1959).

Bibliography

[27] Armitage N.P., "Electrodynamics of correlated electron systems," in *Lecture Notes 2008 Boulder Summer School on condensed matter physics* (2008).

[28] J Hubbard, "Electron correlations in narrow energy bands," Proceedings of the Royal Society of London Series A-Mathematical and Physical Sciences **276**, 238 (1963).

[29] *The Hubbard Model: Its Physics and its Mathematical Physics*, edited by Dionys Baeriswyl, David K. Campbell, Jose M.P. Carmelo, Francisco Guinea, and Enrique Louis (Springer, 1995).

[30] M Dressel and N Drichko, "Optical properties of two-dimensional organic conductors: Signatures of charge ordering and correlation effects," Chemical Reviews **104**, 5689 (2004).

[31] H Kuroda, K Yakushi, H Tajima, A Ugawa, M Tamura, Y Okawa, A Kobayashi, R Kato, H Kobayashi, and G Saito, "Reflectance Spectra of β-(BEDT-TTF$_2$I$_3$, θ-(BEDT-TTF)$_2$I$_3$, and κ-(BEDT-TTF)$_2$I$_3$, and β"-(BEDT-TTF)$_2$ICl$_2$ and β'-(BEDT-TTF)$_2$ICl$_2$ - Relation between the Inter-band Transition and the Dimeric Structure," Synthetic Metals **27**, A491 (1988).

[32] H Tajima, S Kyoden, H Mori, and S Tanaka, "Estimation of charge-ordering patterns in θ-ET$_2$MM '(SCN)$_4$ (MM ' = RbCo, RbZn, CsZn) by reflection spectroscopy," Physical Review B **62**, 9378 (2000).

[33] J Merino and RH McKenzie, "Transport properties of strongly correlated metals: A dynamical mean-field approach," Physical Review B **61**, 7996 (2000).

[34] J Merino and RH McKenzie, "Cyclotron effective masses in layered metals," Physical Review B **62**, 2416 (2000).

[35] J. Wosnitza, "Quasi-two-dimensional organic superconductors," Journal of Low Temperature Physics **146**, 641 (2007).

[36] IJ Bruno, JC Cole, PR Edgington, M Kessler, CF Macrae, P McCabe, J Pearson, and R Taylor, "New software for searching the Cambridge Structural Database and visualizing crystal structures," Acta Crystallographica Section B-Structural Science **58**, 389 (2002).

[37] H. Haken and H. C. Wolf, *Molekülphysik und Quantenchemie* (Springer, 2006).

[38] T Mori, H Mori, and S Tanaka, "Structural genealogy of BEDT-TTF-based organic conductors - II. Inclined molecules: theta, alpha, and chi phases," Bulletin of the Chemical Society of Japan **72**, 179 (1999).

[39] H Mori, N Sakurai, S Tanaka, H Moriyama, T Mori, H Kobayashi, and A Kobayashi, "Control of electronic state by dihedral angle in theta-type bis(ethylenedithio)tetraselenafulvalene salts," Chemistry of Materials **12**, 2984 (2000).

[40] J Yamaura, A Miyazaki, T Enoki, and G Saito, "Crystal structure and magnetic properties of the organic antiferromagnet (C(1)TET-TTF)$_2$Br," Physical Review B **55**, 3649 (1997).

[41] F Mila, "Deducing correlation parameters from optical conductivity in the bechgaard salts," Physical Review B **52**, 4788 (1995).

[42] T Enoki, T Umeyama, A Miyazaki, H Nishikawa, I Ikemoto, and K Kikuchi, "Novel metallic state carrying localized spins in the molecular conductor (DMET)$_2$FeBr$_4$," Physical Review Letters **81**, 3719 (1998).

[43] S Mazumdar, RT Clay, and DK Campbell, "Bond-order and charge-density waves in the isotropic interacting two-dimensional quarter-filled band and the insulating state proximate to organic superconductivity," Physical Review B **62**, 13400 (2000).

[44] RH McKenzie, J Merino, JB Marston, and OP Sushkov, "Charge ordering and antiferromagnetic exchange in layered molecular crystals of the theta type," Physical Review B **64**, 085109 (2001).

[45] CS Jacobsen, JM Williams, and HH Wang, "Infrared properties of the ambient pressure organic superconductor (BEDT-TTF)$_2$I$_3$," Solid State Communications **54**, 937 (1985).

[46] S Horiuchi, H Yamochi, G Saito, K Sakaguchi, and M Kusunoki, "Nature and origin of stable metallic state in organic charge-transfer complexes of bis(ethylenedioxy)tetrathiafulvalene," Journal of the American Chemical Society **118**, 8604 (1996).

[47] G Saito, *Metal-Insulator Transition Revisited* (TAYLOR & FRANCIS LTD, London, 1995).

[48] VM Yartsev, OO Drozdova, VN Semkin, RM Vlasova, and RN Lyubovskaya, "Optical properties of deuterated organic conductor (BEDT-TTF)$_2$[Hg(SCN)$_2$Br]," Physica Status Solidi B-Basic Research **209**, 471 (1998).

[49] A Fortunelli and A Painelli, "Ab initio estimate of hubbard model parameters: A simple procedure applied to BEDT-TTF salts," Physical Review B **55**, 16088 (1997).

[50] JE Eldridge, K Kornelsen, HH Wang, JM Williams, AVS Crouch, and DM Watkins, "Infrared optical properties of the 12-K organic superconductor κ-(BEDT-TTF)$_2$Cu[N(CN)$_2$]Br," Solid State Communications **79**, 583 (1991).

[51] KE Kornelsen, JE Eldridge, HH Wang, and JM Williams, "Far-Infrared Optical-Absorption of the 10.4 K organic superconductor κ-(BEDT-TTF)$_2$[Cu(NCS)$_2$]," Solid State Communications **76**, 1009 (1990).

[52] K Kornelsen, J Eldridge, HH Wang, and JM Williams, "Infrared-Optical Properties of the Deuterated Form of the 10K Organic Superconductor (BEDT-TTF)$_2$[Cu(NCS)$_2$]," Solid State Communications **74**, 501 (1990).

[53] Daniel Faltermeier, Jakob Barz, Michael Dumm, Martin Dressel, Natalia Drichko, Boris Petrov, Victor Semkin, Rema Vlasova, Cecile Meziere, and Patrick Batail, "Bandwidth-controlled mott transition in κ-(BEDT-TTF)$_2$Cu[N(CN)$_2$]Br$_x$Cl$_{1-x}$: Optical studies of localized charge excitations," Physical Review B **76**, 165113 (2007).

[54] H Kino and H Fukuyama, "Phase diagram of two-dimensional organic conductors: (BEDT-TTF)$_2$X," Journal of the Physical Society of Japan **65**, 2158 (1996).

[55] K Kornelsen, JE Eldridge, CC Homes, HH Wang, and JM Williams, "Optical-Properties of the 10-K Organic Superconductor (BEDT-TTF)$_2$[Cu(SCN)$_2$]," Solid State Communications **72**, 475 (1989).

[56] O Gunnarsson, M Calandra, and JE Han, "Colloquium: Saturation of electrical resistivity," Reviews of Modern Physics **75**, 1085 (2003).

[57] G Kotliar, S.Y. Savrasov, K Haule, V.S. Oudovenko, O. Parcollet, and C.A. Marianetti, "Electronic structure calculations with dynamical meas-field theory," Reviews of Modern Physics **78**, 865 (2006).

[58] M Dressel, D Faltermeier, M Dumm, N Drichko, B Petrov, V Semkin, R Vlasova, C Meziere, and P Batail, "Disentangling the conductivity spectra of two-dimensional organic conductors," Physica B: Condensed Matter **404**, 541 (2009).

[59] M Dumm, D Faltermeier, N Drichko, M Dressel, C Meziere, and P Batail, "Bandwidth-controlled Mott transition in κ-(BEDT-TTF)$_2$Cu[N(CN)$_2$]Br$_x$Cl$_{1-x}$: Optical studies of correlated carriers," Physical Review B **79**, 195106 (2009).

[60] J Merino, M Dumm, N Drichko, M Dressel, and RH McKenzie, "Quasiparticles at the verge of localization near the Mott metal-insulator transition in two-dimensional material," Physical Review Letters **100**, 086404 (2008).

[61] M Dressel, G Gruner, JP Pouget, A Breining, and D Schweitzer, "Field and Frequency-Dependent Transport in the 2-Dimensional Organic Conductor α-(BEDT-TTF)$_2$I$_3$," Journal De Physique I **4**, 579 (1994).

[62] T Nakamura, W Minagawa, R Kinami, and T Takahashi, "Possible charge disproportionation and new type charge localization in θ-(BEDT-TTF)$_2$CsZn(SCN)$_4$," Journal of the Physical Society of Japan **69**, 504 (2000).

[63] K Miyagawa, A Kawamoto, and K Kanoda, "Charge ordering in a quasi-two dimensional organic conductor," Physical Review B **62**, R7679 (2000).

[64] Y Takano, K Hiraki, HM Yamamoto, T Nakamura, and T Takahashi, "Charge ordering in α-(BEDT-TTF)$_2$I$_3$," Synthetic Metals **120**, 1081 (2001).

[65] NL Wang, H Mori, S Tanaka, J Dong, and BP Clayman, "Far-infrared study of the insulator-metal transition in θ-(BEDT-TTF)$_2$RbZn(SCN)$_4$ (BEDT-TTF equivalent to bis (ethylene-dithio)tetrathiafulvalene)," Journal of Physics-Condensed Matter **13**, 5463 (2001).

[66] J Dong, JL Musfeldt, JA Schlueter, JM Williams, PG Nixon, RW Winter, and GL Gard, "Optical properties of β"-ET$_2$SF$_5$CH$_2$CF$_2$SO$_3$: A layered molecular superconductor with large discrete counterions," Physical Review B **60**, 4342 (1999).

[67] M Dressel, N Drichko, J Schlueter, and J Merino, "Proximity of the layered organic conductors α-(BEDT-TTF)$_2$MHg(SCN)$_4$ (M=K,NH$_4$) to a charge-ordering transition," Physical Review Letters **90**, 167002 (2003).

[68] H Kino and H Fukuyama, "Interrelationship among electronic states of α-ET$_2$I$_3$, ET$_2$MHg(SCN)$_4$ and κ-ET2X," Journal of the Physical Society of Japan **64**, 4523 (1995).

[69] M Calandra, J Merino, and RH McKenzie, "Metal-insulator transition and charge ordering in the extended Hubbard model at one-quarter filling," Physical Review B **66**, 195102 (2002).

[70] J Moldenhauer, C Horn, KI Pokhodnia, D Schweitzer, I Heinen, and HJ Keller, "FT-IR Absorption-Spectroscopy of BEDT-TTF Radical Salts - Charge-Transfer and Donor Anion Interaction," Synthetic Metals **60**, 31 (1993).

[71] C. Clauss, N. Drichko, D. Schweitzer, and M. Dressel, "Charge-order gap in α-(BEDT-TTF)$_2$I$_3$," Physica B: Condensed Matter, in press(2009).

[72] R Wojciechowski, K Yamamoto, K Yakushi, M Inokuchi, and A Kawamoto, "High-pressure Raman study of the charge ordering in α-(BEDT-TTF)$_2$I$_3$," Physical Review B **67**, 224105 (2003).

[73] R Wojciechowski, K Yamamoto, K Yakushi, and A Kawamoto, "Raman study of charge disproportionation in α-(BEDT-TTF)$_2$I$_3$," Synthetic Metals **135**, 587 (2003).

[74] K Yamamoto, K Yakushi, K Miyagawa, K Kanoda, and A Kawamoto, "Charge ordering in θ-(BEDT-TTF)$_2$RbZn(SCN)$_4$ studied by vibrational spectroscopy," Physical Review B **65**, 085110 (2002).

[75] K Suzuki, K Yamamoto, M Uruichi, and K Yakushi, "Charge ordering in θ-(BEDT-TTF)$_2$TlM(SCN)$_4$ [M=Co, Zn] studied by vibrational spectroscopy," Synthetic Metals **135**, 525 (2003).

[76] T Takahashi, R Chiba, Y Takano, Y Kubo, K Hiraki, H Yamamoto, and T Nakamura, "Charge ordering in non-dimerized BEDT-TTF based organic conductors: C-13-NMR experiments," JOURNAL DE PHYSIQUE IV **12**, 201 (2002).

[77] R Chiba, H Yamamoto, K Hiraki, T Takahashi, and T Nakamura, "Charge disproportionation in (BEDT-TTF)$_2$RbZn(SCN)$_4$," (JAN-FEB 2001).

[78] N. Drichko, M. Dressel, C. A. Kuntscher, A. Pashkin, A. Greco, J. Merino, and J. Schlueter, "Electronic properties of correlated metals in the vicinity of a charge-order transition: Optical spectroscopy of α-(BEDT-TTF)$_2$MHg(SCN)$_4$ (M=NH$_4$, Rb, Tl)," Physical Review B **74**, 235121 (2006).

[79] H Mori, S Tanaka, Oshima M, G Saito, T Mori, Y Maruyama, and H Inokuchi, "Crystal and Electronic-Structures of (BEDT-TTF)$_2$(KHg(SCN)$_4$), (BEDT-TTF)$_2$(NH$_4$Hg(SCN)$_4$)," Bulletin of the Chemical Society of Japan **63**, 2183 (1990).

[80] T Osada, R Yagi, A Kawasumi, S Kagoshima, N Miura, M Oshima, and G Saito, "High-field magnetotransport and Fermi-surface topology in the novel quasi-two-dimensional organic conductor bis(ethylenedithiolo)tetrathiafulvalenium mercuric postassium thiocyanate, (BEDT-TTF)$_2$KHg(SCN)$_4$," Physical Review B **41**, 5428 (1990).

[81] J Singleton, "Studies of quasi-two-dimensional organic conductors based on BEDT-TTF using high magnetic fields," Reports On Progress In Physics **63**, 1111 (2000).

[82] M Maesato, Y Kaga, R Kondo, and S Kagoshima, "Control of electronic properties of α-(BEDT-TTF)$_2$MHg(SCN)$_4$ (M=K,NH$_4$) by the uniaxial strain method," Physical Review B **64**, 155104 (2001).

[83] BH Ward, JA Schlueter, U Geiser, HH Wang, E Morales, JP Parakka, SY Thomas, JM Williams, PG Nixon, RW Winter, GL Gard, HJ Koo, and MH Whangbo, "Comparison of the crystal and electronic structures of three 2 : 1 salts of the organic donor molecule BEDT-TTF with pentafluorothiomethylsulfonate anions SF$_5$CH$_2$SO$_3$-, SF$_5$CHFSO$_3$-, and SF$_5$CF$_2$SO$_3$-," Chemistry of Materials **12**, 343 (2000).

[84] BR Jones, I Olejniczak, J Dong, JM Pigos, ZT Zhu, AD Garlach, JL Musfeldt, HJ Koo, MH Whangbo, JA Schlueter, BH Ward, E Morales, AM Kini, RW Winter, J Mohtasham, and GL Gard, "Optical spectra and electronic band structure calculations of β"-ET$_2$SF$_5$RSO$_3$ (R = CH$_2$CF$_2$, CHFCF$_2$, and CHF): Changing electronic properties by chemical tuning of the counterion," Chemistry of Materials **12**, 2490 (2000).

[85] H.M. Yamamoto, H Mutsumi, Y. Kawasugi, K. Tsukagoshi, and R. kato, "Field effect on organic charge-ordered/mott insulators," Physica B **404**, 413 (2009).

[86] Conrad Clauss, *Optical Investigations on the Charge Ordered State of Two Dimensional Conductors*, Master's thesis, 1. Physikalisches Institut, Universität Stuttgart (2008).

[87] Masashi Watanabe, Kenichiro Yamamoto, Takayoshi Ito, Yusaku Nakashima, Makoto Tanabe, Noriaki Hanasaki, Naoshi Ikeda, Yoshio Nogami, Hiroyuki Ohsumi, Hidenori Toyokawa, Yukio Noda, Ichiro Terasaki, Fumiaki Sawano, Tomohiro Suko, Hatsumi Mori, and Takehiko Mori, "Non-thermal evidence for current-induced melting of charge order in θ-(BEDT-TTF)$_2$CsZn(SCN)$_4$," Journal of the Physical Society of Japan **77**, 065004 (2008).

[88] N Drichko, K Petukhov, M Dressel, O Bogdanova, E Zhilyaeva, R Lyubovskaya, A Greco, and J Merino, "Indications of electronic correlations in the 1/5-filled two-dimensional conductor β ''-(BEDO- TTF)$_5$[CsHg(SCN)$_4$]$_2$," Physical Review B **72**, 024524 (2005).

[89] N Drichko, M Dumm, D Faltermeier, M Dressel, J Merino, and A Greco, "Drude weight in quasi-two-dimensional organic conductors close to the Mott transition: Optical studies of the bandwidth, filling and temperature dependence," Physica C **460**, 125 (2007).

[90] DC Mattic and J Bardeen, "Theory of the anomalous skin effect in normal and superconducting metals," Physical Review **111**, 412 (1958).

[91] W Zimmermann, EH Brandt, M Bauer, E Seider, and L Genzel, "optical Conductivity of BCS Superconductors with Arbitrary Purity," Physica C **183**, 99 (1991).

[92] A Girlando, M Masino, G Visentini, RG Della Valle, A Brillante, and E Venuti, "Lattice dynamics and electron-phonon coupling in the β-(BEDT-TTF)$_2$I$_3$ organic superconductor," Physical Review B **62**, 14476 (2000).

[93] ME Kozlov, KI Pokhodnia, and AA Yurchenko, "The Assignment of Fundamental Vibrations of BEDT-TTF and BEDT-TTF-d8," Spectrochimica Acta Part A-Molecular and Biomolecular Spectroscopy **43**, 323 (1987).

[94] R Bozio and C Pecile, *Spectroscopy of Advanced Materials* (WILEY & SONS, 1991).

[95] M Meneghetti, R Bozio, and C Pecile, "Electron-Molecular Vibration Coupling in 2-D Organic Conductors - High and Low-Temperature Phases of α-(BEDT-TTF)$_2$I$_3$," Journal De Physique **47**, 1377 (1986).

[96] M Meneghetti, R Bozio, and C Pecile, "Infrared Properties of a 2-D Organic Conductor - α-(BEDT-TTF)$_2$I$_3$ in its High and Low-Temperature Phases," Synthetic Metals **19**, 143 (1987).

[97] ME Kozlov, KI Pokhodnia, and AA YurchenkoO, "Electron Molecular Vibration Coupling in Vibrational-Spectra of BEDT-TTF Based Radical Cation Salts," Spectrochimica Acta Part A-Molecular and Biomolecular Spectroscopy **45**, 437 (1989).

[98] A Girlando, M Masino, A Brillante, RG Della Valle, and E Venuti, "BEDT-TTF organic superconductors: The role of phonons," Physical Review B **66**, 100507(R) (2002).

[99] JL Musfeldt, R Swietlik, I Olejniczak, JE Eldridge, and U Geiser, "Understanding electron-molecular vibrational coupling in organic molecular solids: Experimental evidence for strong coupling of the 890 cm^{-1} mode in ET-based materials," Physical Review B **72**, 014516 (2005).

[100] T Yamamoto, M Uruichi, K Yamamoto, K Yakushi, A Kawamoto, and H Taniguchi, "Examination of the charge-sensitive vibrational modes in bis(ethylenedithio)tetrathiafulvalene," Journal of Physical Chemistry B **109**, 15226 (2005).

[101] HH Wang, JR Ferraro, JM Williams, U Geiser, and JA Schlueter, "Rapid Raman-Spectroscopic Determination of the Stoichiometry of Microscopic Quantities of BEDT-TTF-Based Organic Conductors and Superconductors," Journal of the Chemical Society-Chemical Communications **16**, 1893 (1994).

[102] HH Wang, AM Kini, and JM Williams, "Raman characterization of the BEDT-TTF(ClO$_4$)$_2$ salt," Molecular Crystals and Liquid Crystals **284**, 211 (1996).

[103] P Guionneau, CJ Kepert, G Bravic, D Chasseau, MR Truter, M Kurmoo, and P Day, "Determining the charge distribution in BEDT-TTF salts," Synthetic Metals **86** (1997).

[104] M Watanabe, Y Noda, Y Nogami, K Oshima, and H Mori, "The charge ordered state and the possible 2k$_F$-CDW state with Fermi surface nesting in organic conductor θ-(BEDT-TTF)$_2$MM'(SCN)$_4$ (MM '= RbZn, CsCo)," JOURNAL DE PHYSIQUE IV **12**, 231 (2002).

[105] K Suzuki, K Yamamoto, M Uruichi, and K Yakushi, "Charge-ordering in θ-(BEDT-TTF)$_2$MM'(SCN)$_4$ [M = Cs, Rb, TI, M' = Zn, Co]," JOURNAL DE PHYSIQUE IV **114**, 379 (2004).

[106] K Suzuki, K Yamamoto, and K Yakushi, "Charge-ordering transition in orthorhombic and monoclinic single-crystals of θ-(BEDT-TTF)$_2$TlZn(SCN)$_4$ studied by vibrational spectroscopy," Physical Review B **69**, 085114 (2004).

[107] Takashi Yamamoto, Hiroshi M. Yamamoto, Reizo Kato, Mikio Uruichi, Kyuya Yakushi, Hiroki Akutsu, Akane Sato-Akutsu, Atsushi Kawamoto, Scott S. Turner, and Peter Day, "Inhomogeneous site charges at the boundary between the insulating, superconducting, and metallic phases of beta '-type bis-ethylenedithio-tetrathiafulvalene molecular charge-transfer salts," Physical Review B **77**, 205120 (2008).

[108] H.S. Gutowsky and A. Saika, "Dissociation, chemical exchange, and proton magnetic resonance in some aqueous electrolytes," J. Chem. Phys. **21**, 1688 (1953).

[109] MJ Rice, NO Lipari, and S Strassler, "Dimerized organic linear-chain conductors and the unambiguous experimental determination of electron-molecular-vibration coupling constants," Physical Review Letters **39**, 1359 (1977).

[110] VM Yartsev and A Graja, "Electron-intramolecular vibration coupling in charge-transfer salts studied by infrared spectroscopy," International Journal of Modern Physics B **12**, 1643 (1998).

[111] G Visentini, M Masino, C Bellitto, and A Girlando, "Experimental determination of BEDT-TTF+ electron-molecular vibration constants through optical microreflectance," Physical Review B **58**, 9460 (1998).

[112] NO Lipari, CB Duke, R Bozio, A Girlando, C Pecile, and A Padva, "Electron-Molecular-Vibration Coupling in 7,7,8,8-Tetracyano-Para-Quinodimethane (TCNQ)," Chemical Physics Letters **44**, 236 (1976).

[113] U Fano, "Effects of configuration interaction on intensities and phase shifts," Physical Review **124**, 1866 (1961).

[114] JA Schlueter, U Geiser, JM Williams, JD Dudek, ME Kelly, JP Flynn, RR Wilson, HI Zakowicz, PP Sche, D Naumann, T Roy, PG Nixon, RW Winter, and GL Gard, "Rational design of organic superconductors through the use of the large, discrete molecular anions $M(CF_3)_4$- (M=Cu, Ag, Au) and $SO_3CF_2CH_2SF_5$-," Synthetic Metals **85**, 1453 (1997).

[115] U Geiser, JA Schlueter, HH Wang, AM Kini, JM Williams, PP Sche, HI Zakowicz, ML VanZile, JD Dudek, PG Nixon, RW Winter, GL Gard, J Ren, and MH Whangbo, "Superconductivity at 5.2 K in an electron donor radical salt of bis(ethylenedithio)tetrathiafulvalene (BEDT-TTF) with the novel polyfluorinated organic anion $SF_5CH_2CF_2SO_3$ (-)," Journal of the American Chemical Society **118**, 9996 (1996).

[116] JA Schlueter, BH Ward, U Geiser, HH Wang, AM Kini, JP Parakka, E Morales, HJ Koo, MH Whangbo, RW Winter, J Mohtasham, and GL Gard, "Crystal structure, physical properties and electronic structure of a new organic conductor β''-$(BEDT-TTF)_2SF_5CHFCF_2SO_3$," J. Mater. Chem. **11**, 2008 (2001).

[117] I Olejniczak, BR Jones, J Dong, JM Pigos, Z Zhu, AD Garlach, JL Musfeldt, HJ Koo, MH Whangbo, JA Schlueter, BH Ward, E Morales, AM Kini, RW Winter, J Mohtasham, and GL Gard, "Optical studies of the β "-(ET)$_2$SF$_5$RSO$_3$ (R=CH$_2$CF$_2$, CHFCF$_2$ and CHF) system: chemical tuning of the counterion," Synthetic Metals **120**, 785 (2001).

[118] J. A. Schlueter, "private communication," .

[119] MS Nam, A Ardavan, JA Symington, J Singleton, N Harrison, CH Mielke, JA Schlueter, RW Winter, and GL Gard, "Thermal activation between Landau levels in the organic superconductor β "-(BEDT-TTF)$-$2SF$_5$CH$_2$CF$_2$SO$_3$," Physical Review Letters **87**, 117001 (2001).

[120] D Beckmann, S Wanka, J Wosnitza, JA Schlueter, JM Williams, PG Nixon, RW Winter, GL Gard, J Ren, and MH Whangbo, "Characterization of the Fermi surface of the organic superconductor β"-(ET)$_2$SF$_5$CH$_2$CF$_2$SO$_3$ by measurements of Shubnikov-de Haas and angle-dependent magnetoresistance oscillations and by electronic band-structure calculations," European Physical Journal B **1**, 295 (1998).

[121] J Hagel, J Wosnitza, C Pfleiderer, JA Schlueter, J Mohtasham, and GL Gard, "Pressure-induced insulating state in an organic superconductor," Physical Review B **68**, 104504 (2003).

[122] J Wosnitza, G Goll, D Beckmann, S Wanka, JA Schlueter, JM Williams, PG Nixon, RW Winter, and GL Gard, "Organic conductors: Fermi-surface and specific-heat studies," Physica B **246**, 104 (1998).

[123] J Wosnitza, S Wanka, J Hagel, R Haussler, H von Lohneysen, JA Schlueter, U Geiser, PG Nixon, RW Winter, and GL Gard, "Shubnikov-de Haas effect in the superconducting state of an organic superconductor," Physical Review B **62**, 11973 (2000).

[124] MS Nam, SJ Blundell, A Ardavan, JA Symington, and J Singleton, "Fermi surface shape and angle-dependent magnetoresistance oscillations," Journal of Physics-Condensed Matter **13**, 2271 (2001).

[125] J. S. Brooks, V. Williams, E. Choi, D. Graf, M. Tokumoto, S. Uji, F. Zuo, J. Wosnitza, J. A. Schlueter, H. Davis, R. W. Winter, G. L. Gard, and K. Storr, "Fermiology and superconductivity at high magnetic fields in a completely organic cation radical salt," New Journal of Physics **8**, 255 (2006).

[126] JM Schrama, J Singleton, RS Edwards, A Ardavan, E Rzepniewski, R Harris, P Goy, M Gross, J Schlueter, M Kurmoo, and P Days, "Millimetre-wave measurements of the bulk magnetoconductivity of anisotropic metals: application to the organic superconductors κ-(BEDT-TTF)$_2$Cu(NCS)$_2$ and β "-(BEDT-TTF)$_2$SF$_5$CH$_2$CF$_2$SO$_3$ (BEDT-TTF bis(ethylene-dithio)tetrathiafulvalene)," Journal of Physics-Condensed Matter **13**, 2235 (2001).

[127] J Wosnitza, J Hagel, JS Qualls, JS Brooks, E Balthes, D Schweitzer, JA Schlueter, U Geiser, J Mohtasham, RW Winter, and GL Gard, "Coherent versus incoherent interlayer transport in layered metals," Physical Review B **65**, 180506 (2002).

[128] S Wanka, J Hagel, D Beckmann, J Wosnitza, JA Schlueter, JM Williams, PG Nixon, RW Winter, and GL Gard, "Specific heat and critical fields of the organic superconductor β "-(BEDT-TTF)$_2$SF$_5$CH$_2$CF$_2$SO$_3$," Physical Review B **57**, 3084 (1998).

[129] JA Schlueter, AM Kini, BH Ward, U Geiser, HH Wang, J Mohtasham, RW Winter, and GL Gard, "Universal inverse deuterium isotope effect on the T_c of BEDT-TTF-based molecular superconductors," Physica C **351**, 261 (2001).

[130] J Muller, M Lang, F Steglich, JA Schlueter, AM Kini, U Geiser, J Mohtasham, RW Winter, GL Gard, T Sasaki, and N Toyota, "Comparative thermal-expansion study of β "-ET$_2$SF$_5$CH$_2$CF$_2$SO$_3$ and κ-ET$_2$Cu(NCS)$_2$: Uniaxial pressure coefficients of T_c and upper critical fields," Physical Review B **61**, 11739 (2000).

[131] J Wosnitza, J Hagel, O Stockert, C Pfleiderer, JA Schlueter, J Mohtasham, and GL Gard, "Fermi-surface reconstruction close to a pressure-induced metal-insulator transition," JOURNAL DE PHYSIQUE IV **114**, 277 (2004).

[132] J. Hagel, O. Ignatchik, J. Wosnitza, C. Pfleiderer, J. A. Schlueter, H. Davis, R. Winter, and G. L. Gard, "Pressure dependence of the electronic properties of the quasi-two-dimensional organic superconductor β "-ET$_2$SF$_5$CH$_2$CF$_2$SO$_3$," Physica C **460-462**, 639 (2007).

[133] RN Lyubovskaya, EI Zhilyaeva, SA Torunova, GA Mousdis, GC Papavassiliou, JAAJ Perenboom, SI Pesotskii, and RB Lyubovskii, "New ambient pressure organic superconductor with T_c=8.1 K : (EDT-TTF)$_4$Hg$_{3-x}$I$_8$)," JOURNAL DE PHYSIQUE IV **114**, 463 (2004).

[134] Elena I. Zhilyaeva, Andrey Y. Kovalevsky, Rustem B. Lyubovskii, Svetlana A. Torunova, George A. Mousdis, George C. Papavassiliou, and Rimma N. Lyubovskaya, "Anion chain structure controlled Behavior of phase transition in quasi-two-dimensional organic metal (EDT-TTF)$_4$[Hg$_3$I$_8$]$_{1-x}$," Crystal Growth & Design **7**, 2768 (2007).

[135] EI Zhilyaeva, AY Kovalevskyi, SA Torunova, GA Mousdis, RB Lyubovskii, GC Papavassiliou, P Coppens, and RN Lyubovskaya, "Structure and conductivity of unsymmetrical pi-donor ethylenedithiodithiadiselenafulvalene iodomercurate (EDT-DTDSF)$_4$Hg$_3$I$_8$," Synthetic Metals **150**, 245 (2005).

[136] EI Zhilyaeva, SA Torunova, RN Lyubovskaya, GA Mousdis, GC Papavassiliou, JAAJ Perenboom, SI Pesotskii, and RB Lyubovskii, "New ambient pressure organic superconductor with T_c=8.1 K based on unsymmetrical donor molecule,

ethylenedithiotetrathiafulvalene: (EDT-TTF)$_4$Hg$_{3-\delta}$ I$_8$, $\delta \approx 0.1 - 0.2$," Synthetic Metals **140**, 151 (2004).

[137] HH Wang, KD Carlson, U Geiser, WK Kwok, MD Vashon, JE Thompson, NF Larsen, GD McCabe, RS Hulscher, and JM Williams, "A New Ambient-Pressure Organic Superconductor - (BEDT-TTF)$_2$NH$_4$Hg(SCN)$_4$," Physica C **166**, 57 (1990).

[138] H Mori, S Tanaka, Oshima M, G Saito, T Mori, Y Maruyama, and H Inokuchi, "Electrical-Properties and Crystal-Structures of Mercury(II) Thiocyanate Salts Based Upon BEDT-TTF with Li+, K+, NH4+, Rb+, and Cs+," Solid State Communications **74**, 1261 (1990).

[139] T Mori, H Inokuchi, H Mori, S Tanaka, M Oshima, and G Saito, "Thermoelectric-Power of (BEDT-TTF)$_2$MHg(SCN)$_4$ [M=K, Rb, and NH$_4$]," Journal of the Physical Society of Japan **59**, 2624 (1990).

[140] J. Wosnitza, *Fermi Surface of Low-Dimenasional Organic Metals and Superconductors* (Springer-Verlag Berlin Heidelberg, 1996).

[141] H Mori, S Tanaka, and T Mori, "Systematic study of the electronic state in theta-type BEDT-TTF organic conductors by changing the electronic correlation," Physical Review B **57**, 12023 (1998).

[142] R Chiba, K Hiraki, T Takahashi, HM Yamamoto, and T Nakamura, "Extremely slow charge fluctuations in the metallic state of the two-dimensional molecular conductor θ-(BEDT-TTF)$_2$RbZn(SCN)$_4$," Physical Review Letters **93** (2004).

[143] T Takahashi, R Chiba, K Hiraki, HM Yamamoto, and T Nakamura, "Dynamical charge disproportionation in metallic state in θ-(BEDT-TTF)$_2$RbZn(SCN)$_4$," JOURNAL DE PHYSIQUE IV **114**, 269 (2004).

[144] D Schweitzer, E Gogu, I Hennig, T Klutz, and HJ Keller, "Electrochemically Prepared Radical Salts of BEDT TTF - Molecular-Metals and Superconductors," Berichte Der Bunsen-Gesellschaft-Physical Chemistry Chemical Physics **91**, 890 (1987).

[145] S Moroto, K Hiraki, Y Takano, Y Kubo, T Takahashi, HM Yamamoto, and T Nakamura, "Charge disproportionation in the metallic state of α-(BEDT-TTF)$_2$I$_3$," JOURNAL DE PHYSIQUE IV **114**, 339 (2004).

[146] T Takahashi, "C-13-NMR studies of charge ordering in organic conductors," Synthetic Metals **133**, 261 (2003).

[147] Y Kubo, Y Takano, K Hiraki, T Takahashi, HM Yamamoto, and T Nakamura, "C^{13}-NMR studies of the 'narrow gap semiconducting' state of α-(BEDT TTF)$_2$)I$_3$ under pressure," Synthetic Metals **135**, 591 (2003).

Bibliography

[148] Y Takano, K Hiraki, HM Yamamoto, T Nakamura, and T Takahashi, "Charge disproportionation in the organic conductor, α-(BEDT-TTF)$_2$I$_3$," Journal of Physics and Chemistry of Solids **62**, 393 (2001).

[149] Toru Kakiuchi, Yusuke Wakabayashi, Hiroshi Sawa, Toshihiro Takahashi, and Toshikazu Nakamura, "Charge ordering in alpha-(BEDT-TTF)$_2$I$_3$ by synchrotron x-ray diffraction," Journal of the Physical Society of Japan **76**, 113702 (2007).

[150] Yasuhiro Tanaka and Kenji Yonemitsu, "Charge order with structural distortion in organic conductors: Comparison between θ-ET$_2$RbZn(SCN)-4 and α-ET$_2$I$_3$," Journal of the Physical Society of Japan **77**, 034708 (2008).

[151] K. Yakushi, K. Suzuki, K. Yamamoto, T. Yamamoto, and A. Kawamoto, "Unusual electronic state of layered θ-ET$_2$X studied by Raman spectroscopy," Journal of Low Temperature Physics **142**, 659 (2006).

[152] R. Chiba, K. Hiraki, T. Takahashi, H. M. Yamamoto, and T. Nakamura, "Charge disproportionation and dynamics in θ-(BEDT-TTF)$_2$CsZn(SCN)$_4$," Physical Review B **77** (2008).

[153] H.-U. Gremlich and H. Günzler, *IR Spektroskopie: Eine Einführung* (WILEY-V C H, 2003).

[154] H. A. Szymanski, *IR Theory and Practice of Infrared Spectroscopy* (Plenum Press New York, 1964).

[155] E. D. Palik and G. Gosh, *Handbook of Optical Constants of Solids* (Academic Press, 1998).

[156] G Kozlov and A Volkov, "Coherent source submillimeter wave spectroscopy," Millimeter and Submillimeter Wave Spectroscopy of Solids **74**, 51 (1998).

[157] C Allen, F Arams, M Wang, and CC Bradley, "Infrared-to-Millimeter, Broadband, Solid State Bolometer Detectors," Applied Optics **8**, 813 (1969).

[158] DH Martin and E Puplett, "Polarised Interferometric Spectrometry for Millimetre and Submillimetre Spectrum," Infrared Physics **10**, 105 (1970).

[159] Daniel Faltermeier, *Optische Untersuchungen an niedrigdimensionalen organischen Supraleitern*, Master's thesis, Universität Stuttgart (2004).

[160] LI BURAVOV, DN FEDUTIN, and SHCHEGOL.IF, "Mechanism of Conductivity of Well-Conducting Complexes on Basis of Tetracyanquinodimethyl," Soviet Physics Jetp-Ussr **32**, 612 (1971).

[161] IF Shchegolev, "Electric and magnetic properties of linear conducting chains," Physica Status Solidi A-Applied Research **12**, 9 (1972).

[162] HW Helberg and M Dressel, "Investigations of organic conductors by the Schegolev method," Journal De Physique I **6**, 1683 (1996).

[163] O Klein, S Donovan, M Dressel, and G Gruner, "Microwave Cavity Perturbation Technique .1. Principles," International Journal of Infrared and Millimeter Waves **14**, 2423 (1993).

[164] S Donovan, O Klein, M Dressel, K Holczer, and G Gruner, "Microwave Cavity Perturbation Technique .2. Experimental Scheme," International Journal of Infrared and Millimeter Waves **14**, 2459 (1993).

[165] M Dressel, O Klein, S Donovan, and G Gruner, "Microwave Cavity Perturbation Technique .3. Application," International Journal of Infrared and Millimeter Waves **14**, 2489 (1993).

[166] HH Wang, ML VanZile, JA Schlueter, U Geiser, AM Kini, PP Sche, HJ Koo, MH Whangbo, PG Nixon, RW Winter, and GL Gard, "In-plane ESR microwave conductivity measurements and electronic band structure studies of the organic superconductor β "-(BEDT-TTF)$_2$SF$_5$CH$_2$CF$_2$SO$_3$," Journal of Physical Chemistry B **103**, 5493 (1999).

[167] J Merino, A Greco, N Drichko, and M Dressel, "Non-fermi liquid behavior in nearly charge ordered layered metals," Physical Review Letters **96**, 216402 (2006).

[168] T Mori, K Kato, Maruyama Y, H Inokuchi, H Mori, I Hirabayashi, and S Tanaka, "Structural and Physical-Properties of a New Organic Superconductor, (BEDT-TTF)$_4$Pd(CN)$_4$H$_2$O," Solid State Communications **82**, 177 (1992).

[169] H Mori, I Hirabayashi, S Tanaka, T Mori, Y Maruyama, and H Inokuchi, "Superconductivity in (BEDT-TTF)$_4$Pt(CN)$_4$H$_2$O," Solid State Communications **80**, 411 (1991).

[170] R Kondo, T Hasegawa, T Mochida, S Kagoshima, and Y Iwasa, "Donor-acceptor type superconductor, (BETS$_2$Cl$_2$TCNQ)," Chemistry Letters **4**, 333 (1999).

[171] T Mori, K Oshima, H Okuno, K Kato, H Mori, and S Tanaka, "Fermi-Surface in an Organic Metal, Bis(EthyleneDioxy)TetraThiaFulavalene Chloride," Physical Review B **51**, 11110 (1995).

[172] S Kahlich, D Schweitzer, C Rovira, JA Paradis, MH Whangbo, I Heinen, HJ Keller, B Nuber, P Bele, H Brunner, and RP Shibaeva, "Characterization of the Fermi-surface and phase-transitions of (BEDO-TTF)$_2$ReO$_4$.H$_2$O by physical property measurements and electronic band-structure calculations," Zeitschrift Fur Physik B-Condensed Matter **94**, 39 (1994).

[173] K Suzuki, K Yamamoto, K Yakushi, and A Kawamoto, "Infrared and Raman studies of θ-(BEDT-TTF)$_2$CsZn(SCN)$_4$: Comparison with the frozen state of θ-(BEDT-TTF)$_2$RbZn(SCN)$_4$," Journal of the Physical Society of Japan **74**, 2631 (2005).

[174] Mikio Uruichi, Kyuya Yakushi, Hiroshi M. Yamamoto, and Reizo Kato, "Infrared and Raman studies of the charge-ordering phase transition at similar to 170 K in the quarter-filled organic conductor, β ''-(ET)(TCNQ)," Journal of the Physical Society of Japan **75**, 074720 (2006).

[175] E Demiralp and WA Goddard, "Vibrational analysis and isotope shifts of BEDT-TTF donor for organic superconductors," Journal of Physical Chemistry a **102**, 2466 (1998).

[176] M Watanabe, Y Noda, Y Nogami, and H Mori, "Transfer integrals and the spatial pattern of charge ordering in θ-(bedt-ttf)$_2$rbzn(scn)$_4$ at 90 k," Journal of the Physical Society of Japan **73**, 116 (2004).

[177] D. Schweitzer, "private communication," .

[178] S. Ciuchi and S. Fratini, "Signatures of polaronic charge ordering in optical and dc conductivity using dynamical mean field theory," Physical Review B **77** (2008).

[179] A Brillante, RG DellaValle, G Visentini, and A Girlando, "Lattice phonons in neutral BEDT-TTF crystal," Chemical Physics Letters **274**, 478 (1997).

[180] RG Della Valle, A Brillante, G Visentini, and A Girlando, "Structure and phonons of α-ET$_2$I$_3$-crystals," Physica B: Condensed Matter **256**, 195 (1999).

[181] RG Della Valle, A Brillante, G Visentini, and A Girlando, "Lattice dynamics and e-ph coupling in BEDT-TTF superconductors," Synthetic Metals **103**, 2083 (1999).

[182] A Girlando, M Masino, RG Della Valle, A Brillante, and E Venuti, "Organic superconductors: How can we increase the critical temperature," Synthetic Metals **138**, 1273 (2003).

[183] F Venuti, RG Della Valle, L Farina, A Brillante, C Vescovi, and A Girlando, "Temperature dependence of structure and phonons of α- and β-TTF crystals," Physical Chemistry Chemical Physics **3**, 4170 (2001).

[184] S. Iwai, K. Yamamoto, A. Kashiwazaki, F. Hiramatsu, H. Nakaya, Y. Kawakami, K. Yakushi, H. Okamoto, H. Mori, and Y. Nishio, "Photoinduced melting of a stripe-type charge-order and metallic domain formation in a layered BEDT-TTF-based organic salt," Physical Review Letters **98**, 097402 (2007).

[185] N Drichko, P Haas, B Gorshunov, D Schweitzer, and M Dressel, "Evidence of the superconducting energy gap in the optical spectra of α_t-(BEDT-TTF)I$_3$," Europhysics Letters **59**, 774 (2002).

[186] B. Pignon, G. Gruener, V. Ta Phuoc, F. Gervais, C. Marin, and L. Ammor, "Comparative infrared study of optimally doped and underdoped La$_{2-x}$Sr$_x$CuO$_4$ single crystals," Journal of Physics-Condensed Matter **20**, 075230 (2008).

[187] K Takenaka, R Shiozaki, S Okuyama, J Nohara, A Osuka, Y Takayanagi, and S Sugai, "Coherent-to-incoherent crossover in the optical conductivity of La$_{2-x}$Sr$_x$CuO$_4$: Charge dynamics of a bad metal," Physical Review B **65**, 092405 (2002).

[188] H Takagi, B Batlogg, HL Kao, J Kwo, RJ Cava, JJ Krajewski, and WF Peck, "Systematic Evolution of Temperature-Dependent Resistivity in La$_{2-x}$Sr$_x$CuO$_4$," Physical Review Letters **69**, 2975 (1992).

[189] JE Han, E Koch, and O Gunnarsson, "Metal-insulator transitions: Influence of lattice structure, Jahn-Teller effect, and Hund's rule coupling," Physical Review Letters **84**, 1276 (2000).

[190] JE Han, O Gunnarsson, and VH Crespi, "Strong superconductivity with local Jahn-Teller phonons in C-60 solids," Physical Review Letters **90**, 167006 (2003).

[191] K Takenaka, M Tamura, N Tajima, H Takagi, J Nohara, and S Sugai, "Collapse of coherent quasiparticle states in θ-(BEDT-TTF)$_2$I$_3$ observed by optical spectroscopy," Physical Review Letters **95**, 227801 (2005).

[192] P Bruesch, S Staessler, and RH Zeller, "Fluctuations and order in a one-dimensional system. A spectroscopical study of the Peierls transition in K$_2$Pt(CN)$_4$Br$_{0.3}$ · 3(H$_2$O)," Physical Review B **12**, 219 (1975).

[193] N Kida and M Tonouchi, "Spectroscopic evidence for a charge-density-wave condensate in a charge-ordered manganite: Observation of a collective excitation mode in Pr$_{0.7}$Ca$_{0.3}$MnO$_3$ by using THz time-domain spectroscopy," Physical Review B **66**, 024401 (2002).

[194] A. Nucara, P. Maselli, P. Calvani, R. Sopracase, M. Ortolani, G. Gruener, M. Cestelli Guidi, U. Schade, and J. Garcia, "Observation of charge-density-wave excitations in manganites," Physical Review Letters **101**, 066407 (2008).

[195] J. Lloyd-Hughes, D. Prabhakaran, A. T. Boothroyd, and M. B. Johnston, "Low-energy collective dynamics of charge stripes in the doped nickelate La$_{2-x}$Sr$_x$NiO$_{4+\delta}$ observed with optical conductivity measurements," Physical Review B **77**, 195114 (2008).

[196] S Cox, J Singleton, RD McDonald, A Migliori, and PB Littlewood, "Sliding charge-density wave in manganites," Nature Materials **7**, 25 (2008).

[197] M Dumm, S Komiya, Y Ando, and DN Basov, "Anisotropic electromagnetic response of lightly doped $La_{2-x}Sr_xCuO_4$ within the CuO_2 planes," Physical Review Letters **91**, 077004 (2003).

[198] A Lucarelli, S Lupi, M Ortolani, P Calvani, P Maselli, M Capizzi, P Giura, H Eisaki, N Kikugawa, T Fujita, M Fujita, and K Yamada, "Phase diagram of $La_{2-x}Sr_xCuO_4$ probed in the infared: Imprints of charge stripe excitations," Physical Review Letters **90**, 037002 (2003).

[199] H Kitano, R Inoue, T Hanguri, A Maeda, N Motoyama, M Takaba, Kojima K, H Eisaki, and S Uchida, "Microwave and millimeter wave spectroscopy in the slightly hole-doped ladders of $Sr_{14}Cu_{24}O_{41}$," Europhysics Letters **56**, 434 (2001).

[200] B Gorshunov, P Haas, T Room, M Dressel, T Vuletic, B Korin-Hamzic, S Tomic, J Akimitsu, and T Nagata, "Charge-density wave formation in $Sr_{14-x}Ca_xCu_{24}O_{41}$," Physical Review B **66**, 060508 (2002).

[201] T Vuletic, B Korin-Hamzic, S Tomic, B Gorshunov, P Haas, T Room, M Dressel, J Akimitsu, T Sasaki, and T Nagata, "Suppression of the charge-density-wave state in $Sr_{14}Cu_{24}O_{41}$ by calcium doping," Physical Review Letters **90**, 257002 (2003).

[202] T Vuletic, B Korin-Hamzic, T Ivek, S Tomic, B Gorshunov, M Dressel, and J Akimitsu, "The spin-ladder and spin-chain system $(La,YSr,Ca)_{14}Cu_{24}O_{41}$: Electronic phases, charge and spin dynamics," Physics Reports **428**, 169 (2006).

[203] J Shumway, S Chattopadhyay, and S Satpathy, "Electron states and electron-phonon coupling in the BEDT-TTF-based organic superconductors," Physical Review B **53**, 6677 (1996).

[204] Y. Kawakami, S. Iwai, T. Fukatsu, M. Miura, N. Yoneyama, T. Sasaki, and N. Kobayashi, "Optical modulation of effective on-site coulomb energy for the mott transition in an organic dimer insulator," Physical Review Letters **103**, 066403 (2009).

[205] Y Kawakami, H Najaya, S Iwai, N Yoneyama, T Sasaki, and N Kobayashi, "Femtosecond mid-IR pump–probe spectroscopy of photoinduced insulator to metal transition in dimer Mott insulator κ-$(BEDT-TTF)_2X$," Journal of Physics and Chemistry of Solids **69**, 3085 (2008).

[206] SE Barnes, "New Method for Anderson Model," Journal of Physics F-Metal Physics **6**, 1375 (1976).

[207] N Read and DM Newns, "A new functional integral formalism for the degenerate Anderson model," J. Phys. C: Solid State Phys. **16**, L1055 (1983).

[208] P Coleman, "New Approach to the Mixed-Valence Problem," Physical Review B **29**, 3035 (1984).

Acknowledgment

I would like to thank everyone who supported me in this work. First of all Prof. Martin Dressel to supervise me an give me the possibility to work on this fascinating and interesting problem. I am sincerely grateful for his support and discussions during my time at the PI1, for all the freedom he gave me to pursue different projects on my own, and all the possibilities he opened for me. Starting from excellent research opportunities in Stuttgart, but also bringing me into contact with a lot of people working on my topic or related fields. Both on conferences and with collaborations on joined projects. That lead to fruitful discussions and measurements in Stuttgart but also at visits for measurements at laboratories in Tallinn or Parma.

I am especially grateful to Dr. Natalia Drichko for directly supervising me and introducing me into the topic of organic conductors and superconductors, as well as into the techniques of FTIR-spectroscopy and microscopy, sample preparation, and low temperature systems. I am deeply grateful for all the time she invested in and dedicated to me, helping me in the lab at day and night, showing patience with me in long discussions, at writing, and also for being a good friend. Through Natalia I got known and introduced to a lot of valuable colleagues in my field. She and Prof. Dressel guided and encouraged me on my way to become an independent researcher and gave me the freedom to push my own projects.

Prof. Bernhard Keimer I would like to thank for kindly taking over the second referee report on my work.

A special pleasure for me was the work together with Yaxiu Sun and Conrad Clauss during their diploma thesis. They were a great help in experiments, data analysis, construction work, and exchange of ideas. Together with them I could manage to keep the lab running. Not only on our projects but also as support for guests in the lab.

For direct support on measurements at the PI1, I have to thank Dr. Shadi Yasin and Michael Glied who supported and guided me at their setups for the transport measurements. At that point a big thank you goes to Gabi Untereiner for technical help in any situation.

My measurements in the THz range would not have been so successful without Dr. Boris Gorshunov. He introduced me into the BWO-based THz setup and showed me all the little tricks and details to be taken care of for a successful measurement. I also appreciate a lot of collaborations on interesting scientific topics with him.

A special thanks goes to Dr. Thomas Room, Dr. Urmas Nagel, and Dr. Dan Hüvonen for the opportunity to work with them at their THz-setup at the KBFI in Tallinn to measure the temperature dependance of the superconducting gap. The same thanks go to Prof. Alberto Girlando and Dr. Matteo Masino who welcomed me at their lab at Parma university. We successfully conducted the Raman measurements and they also provided lots of fruitful discussions and supporting calculations on charge order in molecular crystals.

All the measurements are not possible without the support of excellent crystals. I have to thank J.A. Schlueter, R.N. Lyubovskaya, H. Mori and D. Schweitzer. They provided me perfect samples, background information and pre-characterization.

On discussion of the results and interpretation of the data I appreciate valuable support by Dr. Michael Dumm, Dr. Neven Barisic, Dr. Andres Greco, and Dr. Jaime Merino. Andres I have to thank for calculations and guidance to understand the theoretical models.

Thank you to all members of the PI1 for the great time you gave me in Stuttgart and the nice working environment. All the support and friendship was valuable and kept the motivation high at all steps of the project.

Finally I have to thank my family, especially my parents, grandparents, and my sister for ongoing and never-ending support and love. Without that this work would not have been possible.

I want morebooks!

Buy your books fast and straightforward online - at one of world's fastest growing online book stores! Environmentally sound due to Print-on-Demand technologies.

Buy your books online at
www.morebooks.shop

Kaufen Sie Ihre Bücher schnell und unkompliziert online – auf einer der am schnellsten wachsenden Buchhandelsplattformen weltweit! Dank Print-On-Demand umwelt- und ressourcenschonend produziert.

Bücher schneller online kaufen
www.morebooks.shop

KS OmniScriptum Publishing
Brivibas gatve 197
LV-1039 Riga, Latvia
Telefax: +371 686 204 55

info@omniscriptum.com
www.omniscriptum.com

Printed by Books on Demand GmbH, Norderstedt / Germany